Contents

Chapter 1
Maize Production: An Overview

Agriculture is the predominant sector in the economy of Jammu and Kashmir. Directly and indirectly, it supports about 80 per cent of the population besides contributing nearly 60 per cent of the state revenue, which adequately explains the over-dependency of the population on agriculture. The overall economic growth of the state depends largely on the progress of the agricultural sector. Jammu and Kashmir is divided into three agro- climatic zones: cold arid desert areas of Ladakh, temperate Kashmir valley and the humid sub-tropical region of Jammu. In the region of Jammu and Kashmir the soils are loamy and there is little clay content in them. Poor in lime but with a high content of magnesia, the soil is treated with chemical fertilizers and enriched with green manure and legume before cultivation. There is sufficient organic matter and nitrogen content in the alluvium of the Kashmir valley as a result of plant residue, crops stubble, natural vegetation and animal excretion. The valley of Kashmir has many types of soils like: Gurti (clay), Bahil (Loam), Sekil (Sandy), Nambaal (Peats), Surzamin, Lemb, Floating garden soils and Karewa soils. No wonder, in Kashmir, soil is virtually worshipped as a miracle of divinity as it is a source of wealth of the land. The cropping patterns of the state vary from region to region due to variations in climate, soil and nature of irrigation, agricultural operations and the system of cultivation.

Jammu region of J&K State is also bestowed with varying climatic conditions and the crops sown in *rabi* are wheat, barley, oats, oilseeds, pulses etc and the crops are sown between mid September – mid January and harvested in May-June in low hills and July-August at high altitude hills, whereas, during *kharif* rice, maize millet, and pulses are sown. Maize is one of the important crops of rainfed agriculture and is grown in low, mid and high hill altitudes. In Kashmir province, land generally produces one crop a year, therefore it is known as *Ekfasli*. There are, of course, exceptions. The highly cultivated garden lands in the neighbourhood of Srinagar and in some other towns give more than one crop in a year. Ploughing for rice, maize and other autumn crops in the Kashmir province commences in the middle of March. In April and May, seeds of these crops are sown. In June and July, barley and wheat, sown in the previous autumn, are harvested. In July and August, linseed is harvested. Cotton picking commences in August and September. Maize, rice and other autumn crops are harvested in September and October. In November and December, ploughing for wheat and barely is undertaken. During the winter months, rice and maize as well as other autumn crops are threshed.

In Ladakh, like Kashmir, no customary rotation of crops is followed. However, wheat is not grown on the same land for more than two or three consecutive years, as this process is believed to weaken the soil. Wheat is always followed by gram. In some villages, land called *Dofasli*, gives two crops a year. In the low-lying areas, where the *kharif* crop maize follows wheat, the former crop is sown anytime from 15 November to 15 January when the soil is not frosty. Maize is sown in July and August. In the villages, where gram is raised as the *rabi* crop instead of wheat, the former is sown immediately after 15 January to give the cultivators sufficient time for growing and harvesting maize in the *kharif*.

1.1 Origin of Maize

It is known that maize had been originated in a wild state in the low lands of Southern Mexico and Central America from which it spread to the Andes. The Amerindians in new Mexico grew it as early as 2,000 B.C. by the time of Columbus maize was growing all the way from great lakes and the lower St. Lawrence valley to Chile and Argentina. In 1492 Columbus discovered cultivated maize in Hiati, where it is known as MAHIZ, a name perhaps originating from Maya peoples responsible for its diffusion. Maize was introduced into Spain by Columbus, but the first attempts of cultivation only took place some 40 years later. During a period in Europe when the cultivation of maize was unknown among the majority of the agriculturists, apart from students and botanists, the Portuguese introduced its use to Guinea and the Congo, from where it has become the staple grain crop for much of sub-Saharan Africa. Maize spread to the rest of the world because of its ability to grow in diverse climates. Maize yields best in warm climate and is now grown in most of the countries that have suitable climatic conditions. Its growth depends more on high summer temperature. A large amount of water is needed during the growth of the maize and its average maturity period is relatively short and this makes it possible to grow at fairly high latitudes. There are about 50 different species of maize having their own characteristic features and kernel size, colour and structure as well as the shape of the kernel, differ from one species to another. Generally, white maize is used for human consumption, and yellow maize is used for corn flakes, animal feeds and industrial products.

1.2 World Maize Production and its Uses

Usually the United States produces nearly one half of the world's supply of maize followed by China, Argentina, Brazil, India, Mexico, South Africa, Italy and Russia. In countries like United States, Australia and Canada maize is known as corn and in India is known as *Bhutta*.

Maize is an annual crop and used for both the animals and humans, it is also known as corn in many countries. It is rich source of carbohydrates, vitamin B, proteins and minerals. It has a nutritional value for both animals and humans, there are so many types of maize like sweet corn, dent corn, flint corn, pop corn, flour corn. Maize has a wide uses that ranges from both human to industrial. Maize is used as a livestock, forage or silage for animals. Humans eat maize or corn in the form of pop corn, porridge, beverage etc. In industries the grains of the maize are used in the transformation of plastic and fabrics. Corn starch is a typical cereal starch used in several industries in many ways, it contains about 66 per cent of starch which is highly digestible for chick, it has good source of vitamins and minerals high in methionine and efficient source of xanthophylls, which valued for their skin and yoke pigmentation in poultry.

The data on world maize production are shown in the following table given below. Production of maize was higher for the United States followed by China, Brazil, Eu-27, Argentina, Mexico and India. India ranks on the 7th position in the world in case of maize production during the year 2007.

Rank Country	*Production (1000MT)*	*Rank Country*	*Production (1000MT)*
United States	272,432.00	China	200,000.00
Brazil	70,000.00	Eu-27	54649.00
Argentina	28,000.00	Ukraine	21,000.00
Mexico	20,700.00	India	20,000.00
South Africa	13,500.00	Canada	11,600.00
Nigeria	9410.00	Indonesia	8900.00
Russia	7500.00	Philippines	7350.00
Egypt	5800.00	Ethiopia	5400.00
Vietnam	5300.00	Thailand	4500.00
Turkey	4,400.00	Serbia	3900.00
Malawi	3600.00	Pakistan	3000.00
Zambia	2850.00	Keniya	2600.00

Source: United States Department of Agriculture, Year of Estimate: 2012.

1.3 Maize Production in Jammu and Kashmir

Maize is the second most important crop of Jammu region, grown on about 67 per cent of the total area and has approximately 84 per cent of state's maize production. The state of Jammu and Kashmir is one of the potential maize growing states, accounting for 4.11 per cent of total maize area and 3.26 per cent of the total national maize area and production, respectively. At national level it ranks VIIIth in the production. The state has 0.32 million hectare under maize cultivation followed by wheat (0.26 million hectare) and rice (0.25 million hectare) with annual production of 0.48 million tones and the average yield of 1.50 tones/ha (Anonymous, 2008). As far as Jammu region of J&K state is concerned, it is a prominent crop and covers about 0.21 million-hectare land, with 0.38 million tonnes

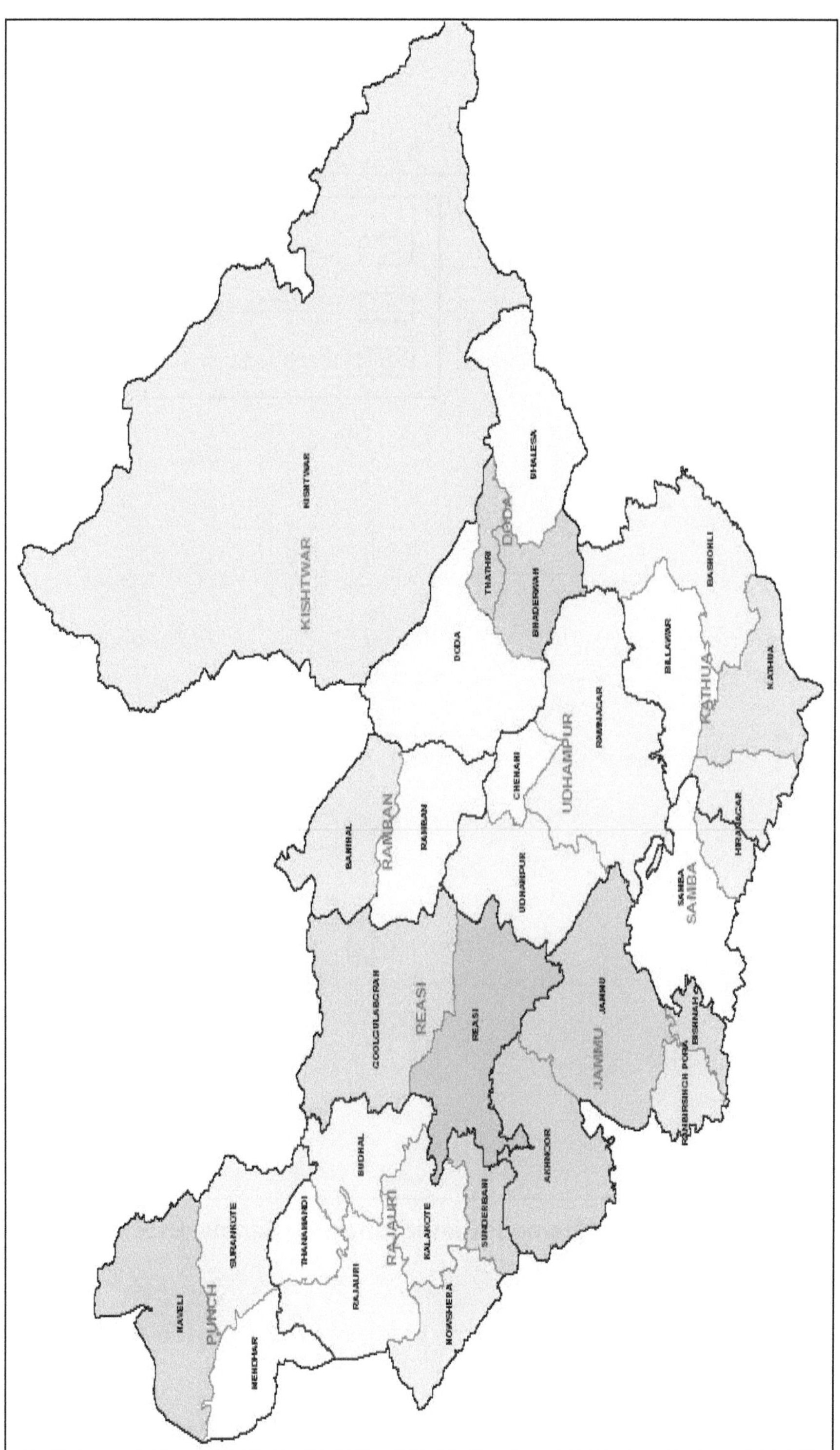

Figure 1.1: Map of Jammu Region.

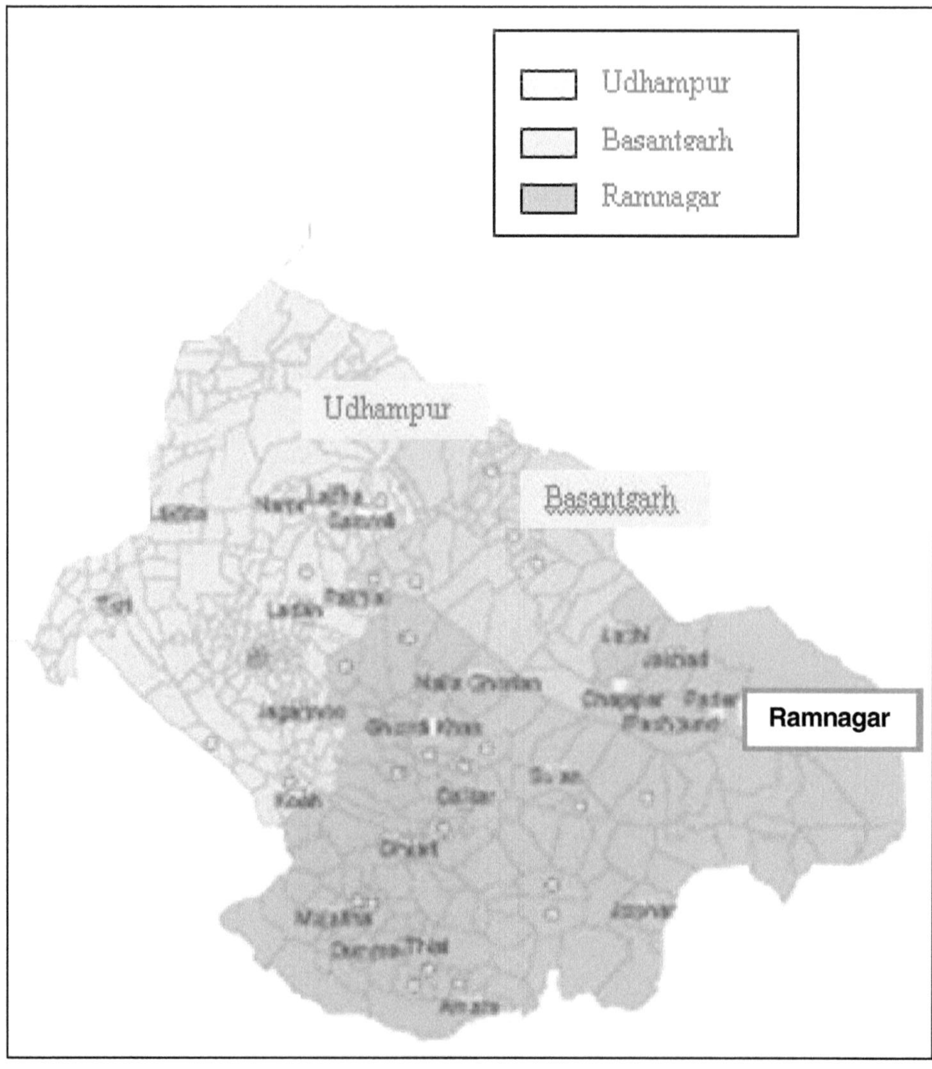

Figure 1.2: Map of Udhampur District showing Sample Blocks.

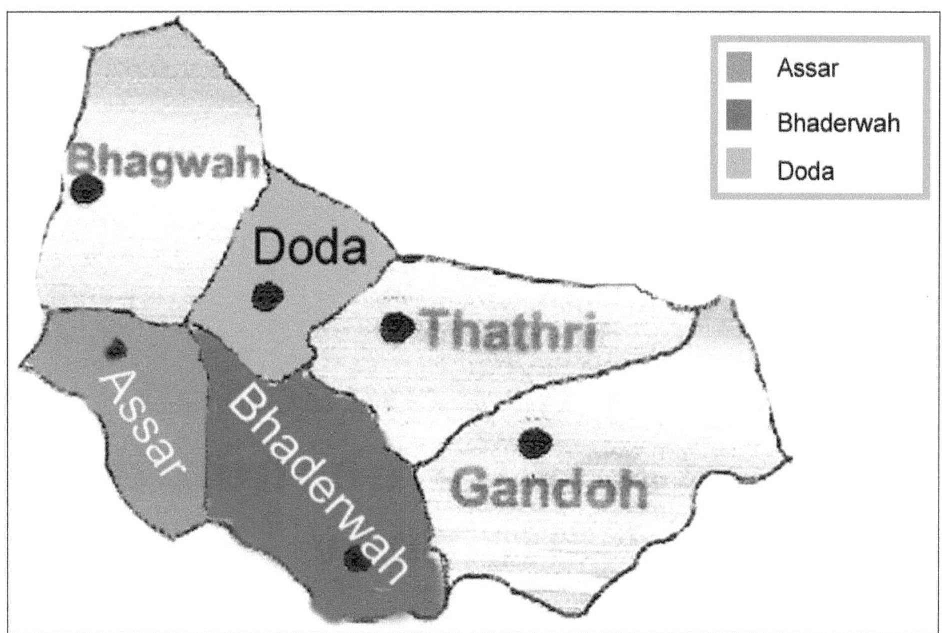

Figure 1.3: Map of Doda District showing Sample Blocks.

of production and 1.73 tonnes/ha productivity, which is marginally lower than the national average of 1.78 tonnes/ha. In Jammu region, maize is grown in six districts of Jammu, Kathua, Udhampur, Rajouri, Poonch, Doda out of total 10 districts. Udhampur and Doda have got the highest area (0.05 and 0.05 million hectares respectively) and production (0.11 and 0.10 million tonnes, respectively). It is an encouraging factor that maize yield of these two districts of Jammu region is more than the national average *i.e.*, 2.12 tonnes/ha for Udhampur and 2.10 tonnes/ha for Doda district.

The two districts of Jammu region, Doda and Udhampur having maximum area under maize cultivation in the Jammu region were selected as shown in the Figures 1.1, 1.2 and 1.3.

Chapter 2
Theory of Production Function and Economic Efficiency

A production function relates to physical output of a production process to physical inputs or factors of production. The production function is one of the key concepts of neoclassical theories, used to define marginal product and to distinguish allocative efficiency, the defining focus of economics. The primary purpose of the production function is to address technical efficiency, resource use efficiency and allocative efficiency in the use of factor inputs in production and the resulting distribution of income to those factors, while abstracting away from the technological problems of achieving technical efficiency, aggregate production functions are estimated to create a framework in which to distinguish how much of economic growth to attribute to changes in factor allocation and how much to attribute to advancing technology.

2.1 Concept of Production Functions

The theory of production is mainly concerned with maximizing profit or in some instances minimizing cost. Both are indicative of economic efficiency. The profit-maximizing firm in perfect competition (taking output and input prices as given) will choose to add input right up to the point where the marginal cost of additional input matches the marginal product in additional output. This implies an ideal division of the income generated from output into an income due to each input factor of production, equal to the marginal product of each input. To satisfy the mathematical definition of a function, a production function is customarily assumed to specify the ***maximum*** output obtainable from a given set of inputs. The production function, therefore, describes a boundary or frontier representing the limit of output obtainable from each feasible combination of input. (Alternatively, a production function can be defined as the specification of the minimum input requirements needed to produce designated quantities of output, given available technology.) By assuming that the maximum output, which is technologically feasible, from a given set of inputs, is obtained, economists are abstracting away from technological, engineering and managerial problems associated with realizing such a technical maximum, to focus exclusively on the problem of allocative efficiency, associated with the ***economic*** choice of how much of a factor input to use, or the degree to which one factor may be substituted for another. In the production function, itself, the relationship of output to input is non-monetary; that is, a production function relates physical inputs to physical outputs, and prices and costs are not reflected in the function.

2.1.1 Economic, Technical and Allocative Efficiency

The theory of production is mainly concerned with maximizing profit or in some instances minimizing cost. Both are indicative of economic efficiency. Farrell (1957) disaggregated economic efficiency into price or allocative efficiency and technical efficiency. Price or allocative efficiency (AE) refers to the marginal conditions for profit maximization. The usual test for allocative efficiency is to compare the MVP of an input to its price. That is, the firm is said to be allocatively efficient when the above condition for profit maximization (*i.e.*, MVP=MFC) is satisfied. Allocative efficiency mainly deals with variable input perspective and equalizes marginal input cost with marginal output price. It guides farmers with managerial decision making about the allocation of variable factors of production- factors that are within the control of firm. Technical efficiency (TE), on the other hand, is related to the fixed responses, which are part of the environment and exogenous of the firm (Yotopoulos and Nugent, 1976). Technical efficiency is the ability of a firm to achieve maximum possible output with available resources. Thus, it is an indicator of productivity of the firm and the variation in TE can reflect the productivity differences among firms. It helps for hunting the potentiality of the existing technologies.

The basic model for explaining the method of measuring technical and allocative efficiencies in the case of one variable input and one output is illustrated in Figure 2.1. The curve TPP_m shows the maximum possible total output as the variable input X increases, while the curve TPP_a shows an average output as the variable input X increases. All points lying below TPP_m are technically inefficient because they give less output at given levels of input. The profit maximization criterion suggests that a producer will choose to utilize X_1 level of input X (where marginal value product of X is equal to its price, P_x) and will produce the technically and allocatively efficient output, Y_1. A producer who uses the input X at X_2 level and produces Y_3 is technically efficient but allocatively inefficient. On the other hand, if he is producing Y_2 by using X_2, he is both technically and allocatively inefficient. Technical efficiency is defined as the ratio of a

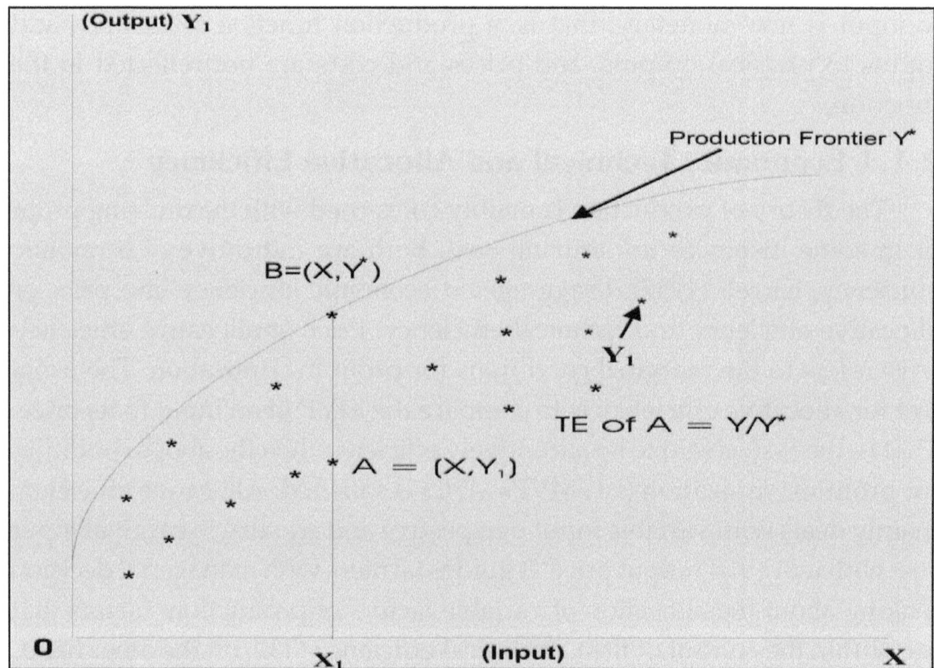

Figure 2.1: Technical Efficiency of Farms in Relative Input-Output Space.

farmer's actual output to the technically maximum possible output at the given level of resources (Y_2/Y_3), allocative efficiency is expressed as the ratio of the technically maximum possible output at the farmer's level of resources to the output obtainable at the optimum level of resources (Y_3/Y_1) and economic efficiency is simply the product of technical and allocative efficiencies $[(Y_2/Y_3)^* (Y_3/Y_1)] = Y_2/Y_1$. Thus, technical, allocative and economic inefficiencies are measured as $(1-Y_2/Y_3)$, $(1-Y_3/Y_1)$ and $(1-Y_2/Y_1)$ respectively (Ali and Chaudhry, 1990).

The production function is defined as the relationship that describes the "maximum possible" output for the given combination of inputs. However, a production function estimated by the ordinary least- squares (OLS) method shows an "average" response (TPP_a) and does not qualify for the theoretical definition of a production frontier (TPP_m). There are a number of approaches used to estimate the technical efficiency, which have been reviewed extensively by Forsund *et al.* (1980). The approaches are summarized in Figure 2.2.

2.1.2 Measurement of Resource Use Efficiency

It is a key factor for increasing productivity. Technical and allocative efficiencies were employed to measure the resource use efficiency. The technical efficiency in production was estimated by using the stochastic frontier production function. The regression co-efficient of factor input from Cobb-Douglas production function (OLS) were used to calculate the Marginal Value Production (MVP) at Geometric mean level for the average farms. In order to study resource – allocative efficiency, the ratio of MVP of a respective input to the marginal factor cost (MFC) for each input was compared and tested for its equality to 1, *i.e.* MVP/MFC= 1 (Yotopoulos, 1967).

If the ratio is equal to one, it means the input is used optimally. If the ratio is more than one, the input is under-utilized and if the ratio is less than one, the input is over-utilized.

2.1.3 Stochastic Production Frontier and Technical Efficiency

For a long time, econometricians have estimated average production functions. It is only after the pioneering work of Farrell (1957) that serious considerations have been given to the possibility of estimating the so-called frontier production functions in an effort to bridge the gap between theory

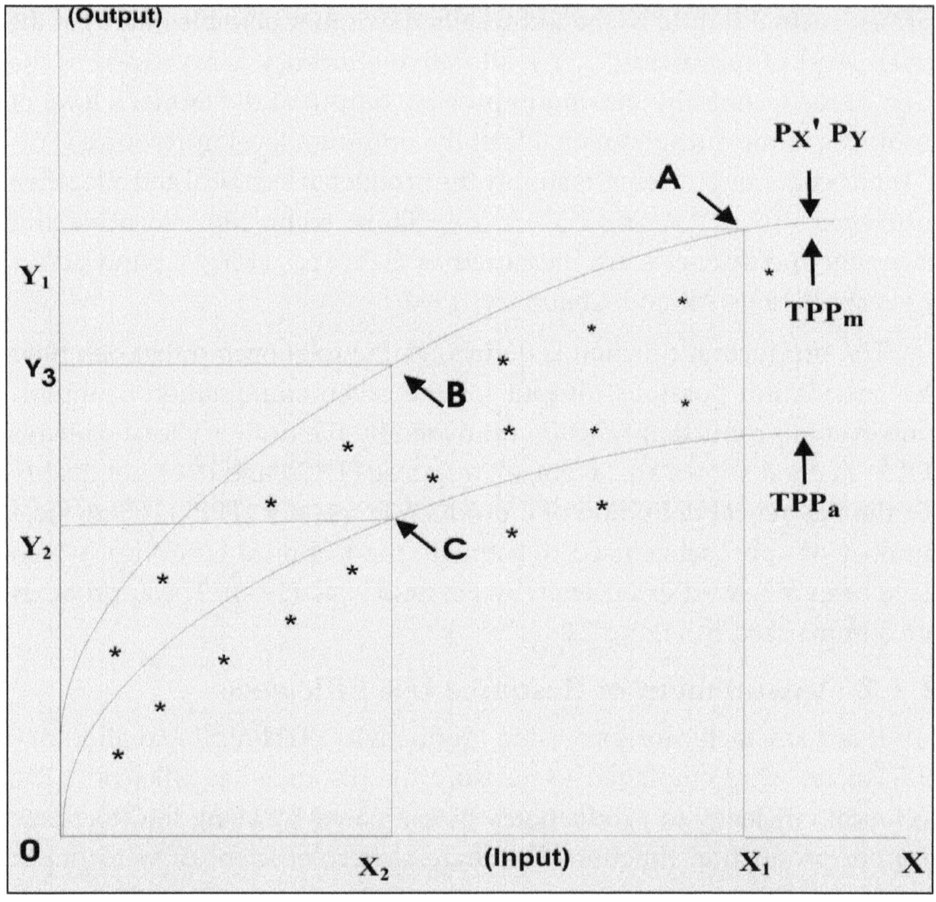

Figure 2.2: Technical, Allocative and Economic Efficiency.

and empirical work. (Aigner, Lovell and Schmidt, 1977). Here, it is important to note that in the traditional production function, socio-economic characteristics and management were not considered as explanatory variables and were thus lumped together in the error term. The stochastic frontier production functions deal with the analysis of socio-economic characteristics of the household that are assumed to be in the composite error term.

The stochastic frontier production function was independently proposed by Aigner, *et al.* (1977), Meeusen and Van den Broeck (1977). This function is defined by;

$$Y_i = f(x_i; \beta) + e_i$$

where,

$i: 1, 2, ..., N$

$e_i: v_i - u_i$

Where Y_i represents the output level of the ith sample farm; $f(x_i; \beta)$ is a suitable function such as Cobb-Douglas or translog production functions of vector, x_i, of inputs for the ith farm and a vector, β, of unknown parameters. e_i is an error term made up of two components: v_i is a random error having zero mean, $N(0; \sigma^2 v,)$ which is associated with random factors such as measurement errors in production and weather over which the farmer does not have control and is assumed to be symmetrical independently distributed as $N(0, \sigma^2 v)$ random variables and independent of u_i. On the other hand, u_i is a non-negative truncated half normal, $N(0, \sigma^2 u)$ random variable associated with farm-specific factors, which lead to the ith firm not attaining maximum efficiency of production; u_i is associated with technical inefficiency of the farm and ranges between zero and one. However, u_i can also have other distributions such as gamma and exponential. N represents the number of farms involved in the cross-sectional survey of the farms. Technical efficiency of an individual farm is defined in terms of the ratio of the observed output to the corresponding frontier output, conditioned on the level of inputs used by the farm. Technical inefficiency is therefore defined as the amount by which the level of production for the farm is less than the frontier output.

$$TE_i\hat{} = Y_i / Y_i^*,$$

where, $Y_i^* = f(x_i; \beta)$, highest predicted value for the ith farm

$$TE_i\hat{} = Exp(-u_i)$$

Technical inefficiency = $1 - TE\hat{}_i$

In their article Bravo-Ureta, *et al.* (1993) suggested that the stochastic frontier production function could be established in two ways. First, if no explicit distribution for the efficiency component is made, and then the production frontier could be estimated using a stochastic version of corrected ordinary least squares (COLS). However, if an explicit distribution is assumed, such as exponential, half-normal or gamma distribution, then the frontier is estimated by maximum likelihood estimates (MLE). According to Greene (1980), MLE makes use of the specific distribution of the disturbance term and this is more efficient than COLS.

The maximum likelihood estimates (MLE) of the parameters of the model defined by equation

$$Y= f (X_i\ \beta) \exp (V_i - U_i).$$

Where Y_i = is the production of the ith farm (i=1, 2, 3- - - - -n), Xi is a (1x k) vector of functions of input quantities applied by the ith farm; β is a (1x k) vector of unknown parameters to be estimated. V_i^s is random variables assumed to be independent. N(O,δ^2v) and independent of U_i^s are the non-negative random variables, associated with technical inefficiency in production assumed to be independently and identically distributed and truncations (at zero) of the normal distribution with mean $Z_i\delta$ and variance σ^2_u (/N(Zi δ, σ^2_u (/); Zi is a (1xm) vector of farm specific variables associated with technical inefficiency, and δ is a(mx1) vector of unknown parameters to be estimated (Battese and Coelli,1995). In the process, the variance parameters σ^2_u and σ^2_v are expressed in terms of the parameterization:

$$\sigma^2 = (\sigma^2_u + \sigma^2_v) \text{ and}$$

$$\gamma = (\sigma^2_u/\sigma^2_u + \sigma^2_v) \text{ or}$$

$$\gamma = \sigma^2_u/\sigma^2$$

In terms of its value and significance, γ is an important parameter in determining the existence of a stochastic frontier. The value of γ ranges between 0 and 1 with the values close to 1 indicating the random component of the inefficiency effects makes a significant contribution to the analysis of the production system (Coelli and Battese, 1996). Similarly, γ = 1 implies that all the deviations from the frontier are entirely due to technical inefficiency (Coelli *et al.*, 1998). Previously, TE was estimated using a two-stage process. First, was to measure the level of efficiency/ inefficiency using a normal production function. Second, was to determine socio-economic characteristics that determine levels of technical efficiency. This was done by using a probit model, with TE as the dependant variable and the socio-economic characteristics as the independent variables. However, since 2000, the stochastic frontier and inefficiency models are jointly estimated using Limdep (Green, 2002) or Frontier computing packages, which apply MLE.

A production function can be expressed in a functional form as the right side of

$$Q = f(X_1, X_2, X_3, \ldots, X_n)$$

where,

Q: Quantity of output

$X_1, X_2, X_3, \ldots, X_n$: Qantities of factor inputs

A linear production function:

$$Q = a + bX_1 + cX_2 + dX_3 + \ldots$$

where, a, b, c and *d* are parameters that are determined empirically.

Another is as a Cobb-Douglas production function:

$$Q = aX_1^b X_2^c \ldots$$

Other forms include the constant elasticity of substitution production function (CES), which is a generalized form of the Cobb-Douglas function, and the quadratic production function. The best form of the equation to use and the values of the parameters (*a, b, c,....*) vary from company to company and industry to industry. In a short run production function at least one of the *X*'s (inputs) is fixed. In the long run all factor inputs are variable at the discretion of management.

2.1.4 Stages of Production

A Production function is divided into 3 stages. In Stage 1 the variable input is being used with increasing output per unit, the latter reaching at a maximum point (since the average physical product is at its maximum at that point). Because the output per unit of the variable input is improving throughout stage 1, a price-taking firm will always operate beyond this stage.

In Stage 2 of production function, output increases at a decreasing rate, and the average and marginal physical product are declining. However, the average product of fixed inputs (not shown) is still rising, because output is rising while fixed input usage is constant. In this stage, the employment of additional variable inputs increases the output per unit of fixed input but decreases the output per unit of the variable input. The optimum input/output combination for the price-taking firm will be in stage 2, although a firm facing a downward-sloped demand curve might find it most profitable to operate in Stage 1. In Stage 3, too much variable input is being used relative to the available fixed inputs: variable inputs are over-utilized in the sense that their presence on the margin obstructs the production process rather than enhancing it. The output per unit of both the fixed and the variable input declines throughout this stage. At the boundary between stage 2 and stage 3, the highest possible output is being obtained from the fixed input.

2.1.5 Shifting a Production Function

In the long run the firm can change its scale of operations by adjusting the level of inputs that are fixed in the short run, thereby shifting the production function upward as plotted against the variable input. Adjustments to the scale of operations may be more significant than what is required to merely balance production capacity with demand.

If a firm is operating at a profit-maximizing level in stage one, it might, in the long run, choose to reduce its scale of operations (by selling capital equipment). By reducing the amount of fixed capital inputs, the production function will shift down. The beginning of stage 2 shifts from B1 to B2. The (unchanged) profit-maximizing output level will now be in stage 2.

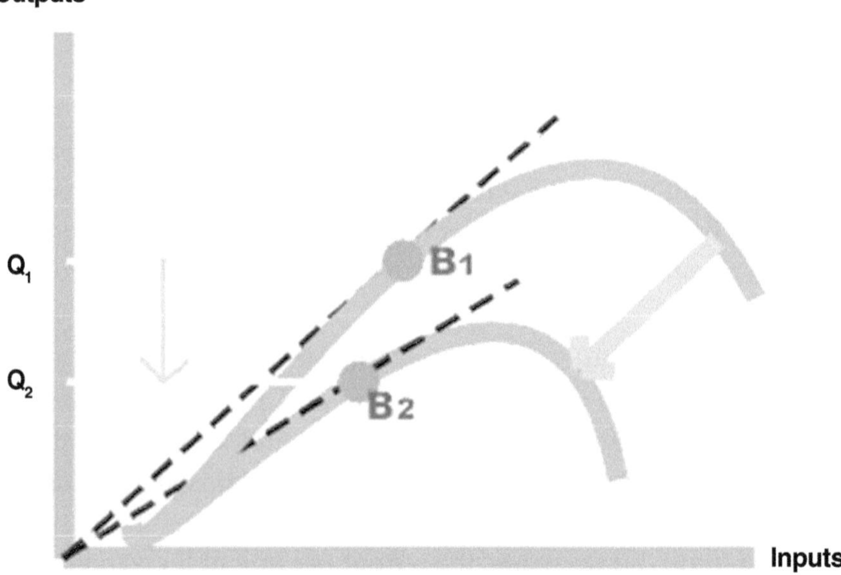

Figure 2.3: Shifting a Production Function (http://en.wikipedia.org/wiki/file)

2.2 Review of Studies Related to Technical Efficiency and its Determinants

The stochastic frontier production function was independently proposed by Aigner, *et al.* (1977) and Meeusen and Van den Broeck (1977) and is used in the estimation of technical efficiency. The technical efficiency of an individual farm is defined in terms of the ratio of the observed output to the corresponding frontier output, conditioned on the level of inputs used by the farm. Technical inefficiency is therefore, defined as the amount by which the level of production for the farm is less than the frontier output.

Kalirajan (1981) estimated a stochastic frontier Cobb-Douglas production function using data from 70 farmers growing IR-20 in Coimbatore district of Tamil Naidu. The productivity differences were mainly related to sample participants individual variability, reflecting differences in farmer's management of variable inputs that was technical efficiency. The major contributing factors to the difference between actual and maximum frontier output among participants were due to limited contact with extension workers, misunderstandings of the farmers and lack of experience and concluded that attention should be focused on extension work, which should be closely integrated with research.

Kalirajan and Flinn (1983) outlined the methodology by which the individual farm effects could be predicted and applied the approach in their analysis of data on 79 rice farmers in the Philippines. A translog stochastic frontier production function was assumed to explain the variation in rice output in terms of several input variables. The parameters of the model were estimated by the method of maximum likelihood estimates. The Cobb-Douglas model was found to be an inadequate model for representation for the farm level data. The individual technical efficiencies ranged from 0.38 to 0.91 per cent. The predicted technical efficiencies were regressed on several farm- level variables and farmer-specific characteristics.

Lingard *et al.* (1983) measured farm-specific technical efficiency of rice farmers in central Luzon used the loop survey data from the International Rice Research Institute (IRRI Philippines) and worked out a production function for 32 farmers from panel data for 1970, 1974, and 1979. They found that measures of technical efficiency, which were calculated from the farm specific dummy variables were least efficient and achieved only 29 per cent of the maximum possible output for given levels.

Ali *et al.* (1987) investigated the technical efficiency of a sample of Illinois grain farms by using a deterministic frontier production function of ray- homothetic frontier, which was estimated by corrected ordinary least squares (COLS) regression and had output and input variables expressed in revenue terms rather than in physical units. Hence, the technical efficiencies also reflected allocative efficiencies. The mean technical efficiency for the 88 grain farms involved was 0.58 in Illinois grain farms and the larger farms were found to be more technically efficient than smaller ones, irrespective of whether acreage cultivated or gross revenue was used to classify the farms by size of operation.

Dawson and Lingard (1989) measured farm- specific technical efficiencies of rice farm in Central Luzon, the Philippines, at discrete points of time. Stochastic frontier production functions were estimated for four years namely 1970, 1974, 1979 and 1982. Results showed that technical inefficiency was the major reason for deviation from the frontier production function. All four samples showed a large range of inefficiency but in general efficiency had improved, particularly between 1979 and 1982.

Ali and Flinn (1989) used a stochastic profit frontier of modified translog type for basmati rice farmers in Pakistan's Punjab, deleted the variables in the translog stochastic profit frontier which had coefficients which were not significantly different from zero individually. After estimating the technical efficiency of individual farmers, the losses in profit due to technical inefficiency were obtained and regressed on various farmer and farm specific variables. Level of education, off- farm employment, unavailability of credit and various constraints associated with irrigation and fertilizer application were significant factors in describing the variability in profit and losses.

Shanmugam and Palanisami (1993) measured the economic efficiency of rice farms using frontier production function commanded under Srivilliputher tank in Kamarajar district of Tamil Nadu. The economic inefficiency revealed that the production could be raised by 29.7 per cent, if the technology gaps between average farmers and best practice farmer were narrowed down by optimum resource allocation in all farms. This highlighted the need for improving the farm extension services to exploit the potential in the available agricultural technologies and it would improve technical efficiency of average farmers.

Kumbhakar (1994) in the study used a flexible translog production function to estimate efficiency of 227 farms from West Bengal. Farm specific technical and allocative inefficiencies were estimated. The mean level of technical efficiency was 75.46 per cent while the best farm was 85.5 per cent efficient (technically). So far as allocative efficiency was concerned the majority of the farms were found to be under users of the endogenous inputs namely, fertilizer, manure, human and bullock labour.

Kalirajan and Shand (1994) measured the technical efficiency of 68 farmers growing High Yielding Variety in the Madurai district of Tamil Nadu and concluded that a substantial variation in the farm specific and input specific elasticity coefficients, which reflected the methods of application of different inputs, varied among sample farms and consequently, individual contributions to inputs and outputs differed from farm to farm. The elasticities of the frontier production function ranged from 0.06 to 0.46 and the efficiency measures ranged from 0.71 per cent to 0.94 per cent. Frequency distribution showed that nearly 50 per cent farmers were in the range of 70–80 per cent of efficiency score.

Bedassa and Krishnamoorthy (1997) used a two-step approach to estimate technical efficiency in paddy farms of Tamil Nadu in India and concluded that the mean technical efficiency was 83.30 per cent, showing potential for increasing paddy production by 17 per cent using present technology. Small and medium-scale farmers were more efficient than the large-scale farms. In addition, the study concluded that animal power was over utilized and therefore suggested its reduction. However, the paddy rice farmers could still benefit by increasing the fertilizer use and expansion of land.

Rajasekharan and Krishnamoorthy (1998) estimated the technical efficiency in rice production and assessed the influence of pesticide on farmers in Anthikkad block of the Kole lands of Kerala using stochastic frontier production function. The farm specific technical efficiencies ranged from 0.49 to 0.92, with a mean of 0.79. Through the help of Logit Model they studied the factors affecting technical efficiency and observed that absence of proper scientific knowledge emerged as one of the determinants of technical efficiency and overuse of pesticides in Kole Lands.

Singh (1999) studied the farm specific technical efficiency of wheat cultivation in Haryana at the aggregate and disaggregate levels by using stochastic frontier approach and reported that wheat-cultivating farms in the state could increase their production by 27 per cent without increasing the quantity of inputs. The estimates of technical efficiency indicated that small sized farms were more efficient than medium and large–sized farms.

Awudu and Richard (2001) used a translog stochastic frontier model to examine technical efficiency in maize and beans in Nicaragua. The average efficiency levels were 69.8 and 74.2 per cent for maize and beans, respectively. In addition, the level of schooling represented human capital, access to formal credit and farming experience (represented by age) contributed positively to production efficiency, while farmers' participation in off-farm employment tended to reduce production efficiency. Large families appeared to be more efficient than small families. Although a larger family size puts extra pressure on farm income for food and clothing, it did ensure availability of enough family labour for farming operations to be performed on time. Positive correlation between inefficiency and participation in non-farm employment suggested that farmers reallocated their time away from farm-related activities, such as adoption of new technologies and gathering of technical information which is essential for enhancing production efficiency. The result indicated that efficiency increased with age until a maximum efficiency was reached when the household head was 38 years old. The age variable probably picked up the effect of physical strength as well as farming experience for the household head.

Shanmugam (2001) attempted the overall technical efficiency and input specific technical efficiency of raising principal crops such as rice, wheat, groundnut and cotton in four major maize growing states of India-Bihar, Karnataka, Punjab and Tamil Nadu using frontier approach and observed that the output of all principal crops selected for the study were less than their respective potential output due to technical inefficiency.

Elsamma and George (2002) assessed the technical efficiency in rice bowl area of Kuttanad village, district Alapuzha (Kerala), using stochastic production function of Cobb-Douglas type. Farm specific technical efficiencies were worked out and found to be in between of 58 per cent to 99 per cent, which indicated that there is a scope to improve farmers' practices through extension and training programmes.

Ahmad *et al.* (2002) estimated technical efficiency of wheat crop by using the stochastic frontier production function and showed that the average technical efficiency of wheat farms was about 68 per cent. An inverse relationship was observed between technical efficiency and farm size besides the farmers with greater access to credit and located closer to the markets were found more efficient. The small farmers were producing at a lower level and farther from the production frontier. The results also revealed that wheat growers in Punjab were comparatively more efficient than their counterparts in Sindh and NWFP.

Shanmugam (2003) measured the farm specific technical efficiency of growing major principal crops in Tamil Nadu, rice, groundnut (irrigated and rainfed), and cotton. It was based on primary data and 600 sample farms were selected by using stratified multistage random sampling method from 60 taluks from Tamil Nadu for the period 1990-91 to 1992-93. The study used the stochastic frontier production function technique to measure the technical efficiency. It indicated that the land, irrigation, labour and fertilizer inputs were the significant determinants of output of almost all crops in the state. The average technical efficiency values of raising selected crops varied from 68 per cent to 82 per cent indicating a scope for rising output without additional resources. It showed that the farmers having larger area were more efficient in cultivating cotton crop. This also indicated that the small farmers can better follow the practices followed by the big farmers to reap more yield or they can shift from cotton cultivation to some other crops for which they are efficient.

Goyal and Suhag (2003) estimated farm level technical inefficiency on wheat farms in Haryana State, India and applied stochastic frontier production function for wheat farmers using unbalanced panel data for three years from 1996-97 to 1998-99. The farm level panel data were collected from 200 farmers and in each year forming 592 total observations. The farm specific technical efficiencies estimated were observed to be time varying. The technical efficiency showed wide variation across sample farms ranging from 0.43 per cent to 0.95 per cent in the 3rd year of the study period. The mean technical efficiency was found to deteriorate through the years in wheat production. It declined from 0.91 per cent in 1st year to 0.90 per cent in 3rd year. The mean technical efficiency indicated that the realized output could be increased by about 10 per cent without any additional resources and more than 2/3rd of total sample farmers had technical efficiency above 0.90 per cent. Various socio-economic and technological factors may be responsible for the observed inefficiency and further rise in inefficiency in wheat production.

Anupama *et al.* (2005) measured technical efficiency in maize and observed that the cost increased significantly on using the improved cultivars due to higher requirements of fertilizers, irrigation and plant protection chemicals, as compared to those in the traditional varieties. However, with the significant increase in yield, the unit cost of production had been much lower in case of improved cultivars. When compared with its competiting crops, the hybrid maize was found superior to soybean as well as paddy in terms of net returns and hence the farmers could cultivate maize rather than paddy, as the later had higher water requirement. They studied that the economic efficiency of the maize growers could be improved by increasing level of the improved package of practices and increased socio-economic status of these farmers to the tune of 23 per cent.

Idiong (2005) estimated the technical efficiency and its determinants using data obtained from 112 small scale swamp rice farmers in Cross River State of Nigeria and observed that mean efficiency obtained was 77 per cent indicating that there was a 23 per cent allowance for improving efficiency. The result also showed that, farmers' educational level, membership in cooperative/farmer association and access to credit significantly influenced the farmers' efficiency positively. He also found that implications of policies would encourage educated persons to form and join cooperatives and provide them with easy access to formal credit.

Ajao *et al.* (2005) examined the technical efficiency of mechanized and non-mechanized maize farmers in Oyo State using stochastic frontier model to access the potentials in maize farms in Nigeria and observed that the mean technical efficiency was 0.72 per cent and 0.62 per cent for mechanized and non-mechanized farms, respectively. It further concluded that the income of respondent could be improved if resources were efficiently used at the existing technology up to 28 per cent. Thus, in the short-run, there was a potential of about 28 per cent to increase the output of maize by adopting the technology and techniques of best practices of maize farms in mechanized farms while the potential therein in non-mechanized farms was about 38 per cent. Also, the entire variable specified in the efficiency model had positive coefficient with fertilizer being the only significant variable for both mechanized and non-mechanized farms.

Ogundari (2006) measured technical efficiency (TE) and allocative efficiency of food and cash crops for rain fed farmers in Nigeria and revealed that herbicide had the highest elasticity followed in order of seeds, fertilizer, land, and labour with least contribution to output. The mean TE index was found to be 0.75 per cent, thereby indicated that 0.25 per cent of rice yield had forgone due to inefficiency. The significant gamma (γ) value of 0.87 established the fact that a high level of technical inefficiency existed among the sampled farmers. The results of the estimation showed that extension contact and access to credit were found to be significant determinants of TE among the farmers and they could adopt appropriate farm practices to prevent under utilization of farm inputs via the intensification of extension activities in the country.

Shehu and Mshelia (2007) estimated the productivity and technical efficiency of 180 small-scale rice farmers in Adamawa State, Nigeria using stochastic frontier production function and the farmers were selected by multi-stage random sampling techniques. The empirical results indicated that the farmers were operating in the irrational stage of production (stage I) as depicted by the returns to scale of 1.06. The predicted technical efficiencies for the farmers ranged from 74 per cent to 98.9 per cent with a mean level of 95.7 per cent. They also observed that an improvement on farmers educational levels through adult education and literacy campaign as well as regulating household size by advocating the need for family planning would probably lead to improvement in technical efficiency in the long term.

Idiong *et al.* (2007) estimated technical efficiency and factors affecting on it in swamp and upland rice farmers in cross river state, Nigeria. A stochastic frontier function that incorporated inefficiency effects was estimated using Maximum Likelihood Estimation (MLE) technique. The results indicated that, swamp and upland rice farmers in the State were not fully technically efficient. They observed that their mean efficiencies were 0.77 per cent and 0.87 per cent for swamp and upland rice systems, respectively and were not significantly different at the 5 per cent level. The results also indicated that farmers educational level, membership of association and access to credit positively influenced their levels of efficiency in both production systems. The coefficients of age and household were negative and significant in determining efficiency of upland rice farmers in the state, while in the swamp they had no significant effect though carried negative signs.

Kamruzzaman and Hedayetul (2008) studied the technical efficiency and factors affecting inefficiency of wheat production in Dinajpur district of Bangladesh. They found that the technical efficiency varied from 40 per cent to 99 per cent and the average was 70.33 per cent. Farmers with optimum sowing and optimum harvesting were technically more efficient than the farmers with late sowing. In all farms technical efficiency was much higher for the farmers who used sandy loam soil for wheat production. They also found that the farmers with educational background and who contacted frequently with extension workers were technically more efficient than those who contacted less with extension workers.

Kachroo *et al.* (2008) studied the measures of the technical efficiency and factors influencing technical efficiency in rice production in Jammu district of Jammu and Kashmir state by using frontier production model. The results revealed that the mean technical efficiency was 37 per cent and the study implied that the average output of rice could be increased by 63 per cent by adopting technology properly and observed that 85 per cent of the observed inefficiency was due to the farmer's decision making and 15 per cent of inefficiency was due to the random factor outside the control of the farmers, the technical efficiency was not much influenced by the proportion of children in the family and dummy for adult members having education above primary level because they had not any significant coefficient.

Oluwatayo *et al.* (2008) examined the technical efficiency and factors which influence the technical efficiency. The stochastic frontier production function was used to determine the technical efficiency of 200 maize growers in rural Nigeria and results presented were based on data collected from a random sample of four villages in the study area using well structured questionnaires. They observed that the estimated technical efficiency of the maize farmers obtained was found to be 0.68 per cent which indicated that 68 per cent efficiency in their use of production inputs and some of the factors found to influence the level of technical efficiency were amount of hired labour, amount of family labour, use of pesticides, use of herbicides and fertilizers application.

Hasan (2008) studied the technical efficiency of maize crop in Sadar Upazila of Dinajpur and Panchagarh in the northern region of Bangladesh and concluded that the farm specific technical efficiency coefficients varied among farmers and ranged from 0.64 per cent to 0.93 per cent with a mean of 0.84 per cent at Danajpur followed by efficiency range 0.52 per cent to 0.94 per cent with mean of 0.80 per cent at Panchagarh. The farms under study were categorized into five different groups with intervals of ten points for better presentation of efficiency results. The results indicated that 30 and 18 per cent of the total farmers at Dinajpur and Panchagarh, respectively belonged to the most efficient category (91 to 100 per cent) and 4 per cent farms at Panchagarh were in the least efficient group (51 to 60 per cent). However, majority of farmers in Dinajpur (42 per cent) belonged to higher efficiency (81 to 90 per cent) group compared to Panchagarh farmers, whose majority belonged to a moderate efficiency (71 to 80 per cent) group.

Ghaderzadeh and Rahimi (2008) conducted a study in the Saqqez city of Kurdistan province, Iran in order to study technical efficiency of wheat farmers. The data were collected from both rainfed and irrigated farms in three different areas (mountain, semi-plain and plain), based on two stage cluster random sampling, for agricultural year 2003-2004. They observed that the variables included in this model were able to explain 66.1 per cent of variation in the average production of rainfed wheat per hectare and estimated transcendental production function model for irrigated farms and showed that the variables included in this model were able to explain 61.8 per cent of variation in the average production of irrigated wheat per hectare. The average technical efficiency for rainfed farms in the mountain, semi-plain and plain areas was 0.67 per cent, 0.65 per cent and 0.65 per cent, respectively. The average technical efficiency for the entire rainfed farms was 0.66 per cent. This showed that rainfed wheat farmers were technically less efficient. In irrigated farms, the average of technical efficiency in mountain, semi-plain and plain areas was 0.70 per cent, 0.68 per cent and 0.65 per cent, respectively and for the entire area was 0.68 per cent. This showed that irrigated wheat farmers in the plain area were technically medium efficient while as in other areas they were technically less efficient.

Javed *et al.* (2008) estimated technical, allocative and economic efficiency to investigate the determinants of technical, allocative and economic inefficiency of rice-wheat cropping system in Punjab, Pakistan and observed that the mean technical, allocative and economic efficiency scores of the sample farms were 0.83 per cent, 0.47 per cent and 0.40 per cent, respectively. An econometric analysis based on Tobit Regression model indicated that farm size, age of farm operator, years of schooling, number of contacts with extension agents, access to credit and farm to market distance were significant determinants of technical efficiency whereas years of schooling, number of contacts with extension agents and access to credit had significant impact on allocative and economic efficiencies of sample farms in the rice-wheat system in Punjab.

Oyewo *et al.* (2009) examined efficiency among 30 maize growing farmers selected through multistage sampling technique in Ogbomoso South local Government in Oyo State by using stochastic frontier production model and pointed out that the quality of maize seed contribution was positive and statistically significant at 1 per cent level. The value of gamma parameter was raised to 0.13, thereby indicate that 13 per cent of the total variation in maize output was due to the technical inefficiencies and the mean technical efficiency was 0.84 and return to scale was 2.77. They further concluded that there was a positive and significant relationship between farm size, quality of seed used and maize output in the study area.

2.3 Review of Resource Use Efficiency Studies

Resource use efficiency in agriculture means that proportion of the total stock that man can make available under technological and economic conditions for cultivation practices. The quantum of production under a normal crop season is directly related to the availability of resource input and their techniques of application. Productivity, the output flow per unit of resource input, depends on the level of input used, which, in turn depends upon the investment pattern (Sharma and Moorti, 1989).

Sahota (1968) conducted study on efficiency of resource allocation in Indian agriculture and opined that on the basis of production function analysis of farm management, the explanatory variables mainly land, human and bullock labour, fixed capital, seed, fertilizers + manure and irrigation produced significant efficiencies of resource allocation in Indian agriculture.

Singh (1971) used Cobb-Douglas production function in his study on resource adjustment possibility and resource use efficiency in the Punjab agriculture to the farm data for various categories of farms pertaining to bet and non-bet regions of Punjab. The coefficients of multiple determination in all regressions were found to be highly significant. Decreasing returns to scale were observed on the tractor operated farms and bullock operated Persian wheel plus canal irrigated farms, Whereas constant return to scale were found on the bullock operated tube well plus canal irrigation. The marginal value productivity was lowest for land on bullock operated farms and highest on tractor operated farms in non-bet area but it was quite contrary in that of bet areas. Marginal rates of substitution of fertilizer for land were higher on canal-irrigated farms and the productivities of land and labour were higher on tractor-operated farms.

Salik and Gupta (1978) examined resource productivity on paddy farms in Chandauli block of Varanasi district. Cobb-Douglas function was fitted to examine the resource productivity and efficiency of resources on farms. It was found that the coefficients of all the resources (area, human labour, bullock labour and value of manure and fertilizer) were significant, there by indicating that using more of these resources could increase the paddy yield. The marginal value products of all the inputs were quite high on adopter farms than on non-adopter farms (farmers with 20 per cent or more area under high yielding varieties of rice were considered as adopters and others were treated as non-adopters).

Alam (1989) studied the resource use efficiency of irrigated and non-irrigated farms in a selected area of Bangladesh from the 78 sampled paddy farmers. On the basis of Cobb-Douglas production function it was concluded that even under irrigated condition land an important factor of production was underpaid, whereas inputs like human labour and animal labour were overpaid. The difference between relative shares and production elasticities of the other factors decreased after recommended irrigation practices were introduced.

Sabur and Haque (1992) conducted a study on resource use efficiency and returns from some selected winter crops in Bangladesh. The study examined the profitability and resource use efficiency of four winter crops (potatoes, wheat, pulses and mustard) grown in Bangladesh from a sample of 200 farmers in Minishganj and Bogra district during the *rabi* season. Potatoes were found to be the most profitable crop in spite of the fact that the production cost for potatoes were also highest of all the crops studied. Therefore, arrangements needed to be made to provide credit that would enable farmers to grow more potatoes. The resource use efficiency was examined by comparing the estimated marginal value of products of inputs with their respective factor cost, which indicated that the farmers were not using the available resources efficiently and concluded that the scope for augmenting profit through optimal allocation of resources needed to be explored.

Singh *et al.* (1992) under unirrigated (96 sample holding) and irrigated (85 sample holding) wheat in the watershed areas of kandi region of lower Shivaliks observed that the wheat yield estimated was 868 kg/ha under unirrigated areas and 1977 kg/ha for irrigated areas, but, fertilizer use (mostly N) was 22.71 kg and 73.87 kg N/ha, respectively, as compared to recommended doses of 70 kg and 250kg N/ha, respectively. He further revealed that next to water, adequate nutrient supply was the major limiting factor for increasing wheat production. The important variables, *viz.* crop area, human labour, bullock labour and fertilizer, were regressed with linear model against the gross returns from wheat. The coefficients indicated that 57 and 79 per cent of the variation in the value of output from the unirrigated and irrigated wheat respectively was explained by these variables. The coefficients were positive and significant for area and fertilizer both for unirrigated and irrigated wheat. The fertilizer response was much higher for unirrigated wheat than for the irrigated indicating that even under rain fed condition fertilizer application was not only essential to get highest yield but also found to be highly remunerative too.

Jha (1996) studied the resource use efficiency on mixed farms from 100 sample size of Kurukshetra district and observed that there was sufficient scope for reconstructing the existing wheat-paddy based cropping system with basmati paddy, potato, sugarcane, sunflower, and summer pulse. However, profitable levels of these crops were restricted due to market imperfections prevalent in the area.

Badal and Singh (2001) examined the resource use efficiency in maize and its competing crops in *rabi* and *kharif* seasons. The primary data of 180 farmers collected during 1996-97 from three districts of Bihar, namely Samastipur, Vaishali and Hazaribagh using the Cobb-Douglas production function. The results showed that HYVs of *rabi* maize offered a greater scope for use of scarce inputs for an enhanced productivity compared to that in the other crops. Human labour used could increase on HYVs of maize and wheat farms both in *rabi* and *kharif* season. Surplus farm labour on local variety maize farms and paddy farms could be employed in some off-farm activities and the earning made available could be used for investment in irrigation and fertilizer use on HYVs of maize farms in *rabi* season.

Singh (2001) conducted a survey of 377 tribal farmers in the Plateau region of Bihar, in order to examine the level of farm resources used in maize and ragi production and to suggest the measures for optimum utilization of these resources. Per hectare yields of maize and ragi on the sample farms were quite high and comparable with the average yields of the state. It was concluded that tribal farmers could increase productivity of maize and ragi on their farms by reallocating their farm resources.

Jyothirmai *et al.* (2003) examined the resource use efficiency of paddy in three regions of Nagarjuna Sagar command area in Andhra Pradesh, where a sample of 120 farmers was selected by random sampling technique. Analysis of resource use efficiency of paddy farms indicated 56 to 61 per cent realization of the maximum possible income by the farmers was from their given set of resources. The analysis further revealed that the efficiency of resource use were low in paddy farms of the command area.

Kumar and Kumar (2004) examined resource use efficiency and returns from selected food grain crops of Himachal Pradesh. The data pertaining to the year 2001-02 were collected by survey method from 200 farms. The results indicated that there existed a vast scope to increase foodgrain production and to turn out negative returns into positive by overcoming the inefficient use of different inputs used in the cultivation of different crops *viz.*, maize, paddy and wheat. The overall analysis of these crops suggested that the farmers should use more of high yielding variety seeds (HYVs) in place of home grown seeds, insecticides and pesticides, bullock labour and tractorization, improved implements etc. in order to increase the agricultural production.

Suresh and Reddy (2006) analyzed resource use efficiency of paddy cultivation in Peechi command area of Thrissur district of Kerela state, where 72 paddy cultivators were selected and Cobb-Douglas production function was used to find out the productivity of resources used in paddy cultivation. It was observed that resource productivity of input used in the cultivation of paddy indicated that the area under paddy, human labour, fertilizer and dummy variables for supplementary irrigation in the case of water stress days were statistically significant. The area under paddy, human labour and fertilizers applied in the cultivation had significant positive elasticity coefficients of 0.65, 0.55 and 0.17, respectively, indicating that at current level of inputs, which were underutilized.

Goni *et al.* (2007) examined the resource use efficiency in rice production in the Lake Chad area of Borno State, Nigeria. Data for the study was collected from 100 rice farmers in four villages and used multi-stage random sampling. Production function analysis was used as the analytical technique. They revealed that the farmers were inefficient in the use of all the resources. Generally, inputs such as seed, land and fertilizer were underutilized. The results showed that there was need for making inputs such as fertilizer and improved seeds affordable and accessible to the farmers so as to improve efficiency. Also policies that encourage the creation of alternative employment opportunities to absorb the excess labour used in rice production in the area should be formulated.

Mohiuddin *et al.* (2007) examined resource use efficiency of maize crop in sadar upazila of Kishoreganj district of Bangladesh. Cobb-Douglas production function was specified to determine the possible relationships between the production of maize and inputs used. They observed that the Cobb-Douglas production function was best fit as indicated by significant F- values and highest R^2 values. The coefficient of multiple determinations (R^2) was 0.82 which means that the explanatory variables obtained explained 82 per cent variation in maize production. The F values were significant at 1 per cent level of confidence indicated that the variation in maize production depended mainly upon the explanatory variables included in the model. The relative contribution of specified factors affecting productivity of maize could be seen from the estimates of regression equation. The results showed that some of the coefficients were not of expected signs. However, the coefficient for quantity of seeds used, irrigation, phosphorus, human labour and land preparation were found to be significant. The values of the coefficients implied that these inputs had considerable effect on return from maize production and the effect was statistically significant.

Kachroo and Kachroo (2007) examined the resource use efficiency of basmati rice in R.S Pura block of Jammu region using the Cobb-Douglas production function. They observed that the model used was best fit to the data as indicated by the highest and significant F- value (11515.80) and R^2 value (0.89), indicated that the model explain 89 per cent of the variation in basmati rice production by the explanatory variables under study. The production coefficients of the variables mainly area and machine labour were negative but significant at 1 per cent level of probability and the plant protection, fertilizers and irrigation were positive and significant means that these inputs have significant influence on the basmati rice production in the area. They further observed that the estimated ratios of MVPs to respective factor cost of variables like irrigation and plant protection measures were more than one, varying from 9.71 for irrigation to 68.15 for fertilizers and for area, machine labour, human labour and seed rate showed the negative values.

Oluwatayo *et al.* (2008) examined resource use efficiency among maize farmers in rural Nigeria. Regression analysis was used to determine the factors that affect maize output in the selected villages. The variables were fitted into four functional forms: linear, semi-log, Cobb-Douglas and exponential forms. The variables tested included farm size, use of tractors, number of days for which tractor was used per season, use of pesticides, use of herbicides, fertilizers, amount of hired and family labour used. The results indicated that the value of R^2 was 93.0 per cent and the regression coefficients for farm size, labour, pesticides, herbicides and fertilizers usage were positively related with output and these variables were equally significant in determining the output of the farmers. They also observed that the farmers who used fertilizers were found to obtain higher yield than those who did not use.

Chapter 3

Production Status of Maize

Maize is one of the most important food grains in the world as well as in developing countries. It is a high yielding crop. Maize occupies first position among the cereals in terms of yield and also gaining significant importance in Indian economy. Globally, maize has great potential as animal feed, human food and industrial end uses, and therefore, emerged as the third most important crop after rice and wheat. Its magnitude of usage in the form of human food, animal and poultry feed and as industrial end uses has given it the much-needed impetus for growth and demand, and therefore was included as component of Integrated Cereals Development Programme (ICDP) and coarse cereals (CC) based cropping systems, before announcement of technology mission on maize in May, 1995. "Accelerated Maize Development Programme (AMDP)" under the technology mission on maize has also introduced in about 26 states of country including J&K State in order to promote its production and productivity. Innovations in the maize sector have resulted in yield gains *i.e.*, approximately 74 per cent over a period of 30 years from 1978-2008 (Anonymous, 2008). Though it is a boon for a country like India because it contributes significantly to India's GDP (₹60 billion annually) and generates huge employment *i.e.*, 450 million man-days (Singh *et al.*, 2003), yet there is a need to pragmatise its strategies for improvement in terms of production, productivity and marketability to bring it at par with world productivity of 4.97 tonnes/ha. Even its growth has faltered in many regions of the country. Compound growth rates and instability for area, production and yield of maize were worked out to know the status of maize in India, Jammu and Kashmir State, Jammu region and selected districts.

3.1 Review of Growth Studies

Arya and Rawat (1990) studied the agricultural growth in Haryana over the period 1966-67 to 1980-81 for state as a whole and for seven districts in existence at the inception of state in 1966. The results indicated that despite inter-district variation, the growth rates in area, production and productivity were found to be negative for oilseeds, sugarcane and cotton in certain districts, whereas area, production and yield growth rate of wheat, rice and potato in all the districts registered an increasing trend. Moreover, barley, millet, sorghum, maize and pulses had registered a declining trend.

Mrigendra and Neelkanth (1996) analyzed the growth rates of productivity of major crops in Madhya Pradesh. The crops taken up for the study included rice, wheat jowar, gram and cotton. The study revealed maximum output growth was in wheat (2.69 per cent per annum) followed by gram (1.33 per cent per annum). The production and productivity of all crops increased during this period with sharp yearly fluctuations. The production of rice varied between 0.18 million tonnes and 0. 57 million tonnes and the variation in production were mainly due to variation in productivity which varied between 383 kg per hectare to 1144 kg per hectare.

Awasthi (2003) estimated compound growth rates for area, production and yield of maize during the three periods *viz.*, 1956-57 to 1965-66, 1966-67 to 1985-86 and 1986-87 to 1998-99 and revealed that during the first period (1956-57 to 1965-66), maize crop registered a growth of 1.32 per cent in area, 11.52 per cent in production and 10.09 per cent in yield per annum at the state level. District wise growth rates indicated that there was considerable variation in the performance of different districts. During the second period (1966-67 to 1985-86) the area, production and yield of maize experienced positive growth rates across all districts with overall state level growth rate of 2.37 per cent, 1.61 per cent and 4.01 per cent for area, yield and production, respectively. He further observed that growth rate in area, production and yield of maize during third period (1986-87 to 1998-99) had declined in comparison to the previous periods and was (-) 0.044 per cent, 1.23 per cent and 1.28 per cent, respectively.

Atibudhi (2003) worked out compound growth rate of maize crop in Orissa and observed that the rate of growth in area, yield and production for the state was 6.53 per cent, 1.66 per cent and 7.86 per cent, respectively for the period 1966-67 to 1985-86. However, during 1986-87 to 1998-99, there had been negative growth in maize area in the state but the yield and production maintained positive growth of 1.92 per cent and 1.88 per cent, respectively. Further, during 1985-86 to 1998-99, there was an increase in production, where as area effect was more than 100 per cent and yield effect was 95.36 per cent for the state.

Varghese and Rathore (2003) studied the growth rate of maize production in the state of Rajasthan and estimated the compound growth rate in area, production and yield in the three temporal phases (i) 1956-57 to 1968-69 (ii) 1969-70 to 1984-85 (iii) 1985–86 to 1999-2000. They reported that during first temporal phase, area and production emerged with positive growth rates in all the districts and yield recorded negative growth in Banswara, Durgapur and Sirohi. In the second temporal phase, the area recorded negative growth in Ajmer and Dungarpur. Production and yield recorded positive growth in all the districts. The growth rate of area was negative in Ajmer, Jhalawar and Udaipur and production was negative in Jhalawar.

Singh *et al.* (2004) examined the growth rates of area, production and productivity of wheat crop in Punjab and reported that the productivity increased by more than five times in five decades from 901 Kg/ha in 1950-51 to 4,696 Kg/ha in 1999-2000. The increase was consistent, continuous and significant. The area under wheat almost increased by three times in the first three decades and was 3 million hectare in 1982-83 whereas in 1999-2000, it was 3.38 million hectare and total production increased by fifteen times in fifty years. Thus, concluded that wheat contributed more than 60 per cent of the states food grains production. However, between 1950-51 and 1999-2000, the annual growth rate of area, yield and production of wheat was 2.25 per cent, 3.43 per cent and 5.46 per cent, respectively.

Mohiuddin *et al.* (2007) studied the growth rates of area, production and yield of maize. For three periods; *i.e.* first period (1980-81 to 1986-87) before the release of the composite varieties and hybrids of maize, second period from the beginning of the released varieties up to the study period (1987-88 to 2000-01) and third period for the overall period (1980-81 to 2000-01). During first seven years from 1980-81 to 1986-87, the annual rate of change of area, production and yield of maize were 11.29, 17.79 and 6.50 per cent, respectively. After the release of composite varieties and hybrids of maize, *i.e.*, after 1986-87, the average area, production and yield of maize increased dramatically and the rate of change were 18.28, 35.18 and 16.91 per cent, respectively.

Hasan (2008) worked out the growth rate of area, production and yield of maize in Bangladesh during 1970-71 to 2005-06 and observed that during first 16 years from 1971-72 to 1986-87, the annual rate of change of area, production and yield of maize were 0.257, 0.049 and 0.108 per cent, respectively. He also observed that after the release of composite varieties and hybrids of maize *i.e.*, after 1986-87, the average area, production and yield of maize increased sharply and the rate of change were 0.985, 0.978 and 0.834 per cent, respectively. The respective mean and coefficient of variation values also support such findings.

3.2 Review of Instability Studies

Instability, in general aims for estimating variability. Variation in area under a crop occurs mainly in response to distribution, timeliness and variations in rainfall and other climatic factors, expected prices and availability of crop-specific inputs. All these factors also affect the variations in yield. Further, yield is also affected by outbreak of diseases, pests, and other natural or man-made hazards like floods, droughts and fire and many other factors. Different events may affect area and yield in the same, opposite or different way.

Jayadevan (1991) studied instability in wheat production in Madhya Pradesh and estimated that the coefficient of variation in wheat production increased in all the individual regions (except Western MP) in the State. During 1965 -66 to 1985-86 yield factor became dominant in all the regions of MP and the coefficient of variation in area under wheat was around 13 per cent in all individual regions except Northern MP during 1951-52 to 1964-65. He also observed that the instability of wheat yield increased in all the individual regions and in the State during 1965-66 to 1986-87 and the area was the main source of production instability in all the individual regions (except Western MP) in 1951-52 to 1964-65.

Radha and Prasad (1999) studied the instability of area, production and productivity of rice and maize in Northern Telangana zone of Andhra Pradesh. Secondary data on area, production and productivity of rice and maize was collected for five districts *viz.*, Nizamabad, Warangal, Khammam, Karimnagar and Adilabad over a period of twenty years, which was divided into two periods to study the changes as well as stability in area, production and yield before and after the implementation of NARP (National Agricultural Research Project) in the zone. The study revealed that the mean values of production and yield exhibited a positive change. Though the coefficient of variation was found to be increasing in area and yield, the production variation decreased *i.e.*, attainment of stability in maize production was observed after the implementation of NARP.

Atibudhi (2003) examined the variability of maize area, yield and production in maize growing districts of Orissa state during 1966-67 to 1985-86. The variability in area was highest in Ganjam followed by Dhenkanal and Korapur districts and variability in yield was highest in Phulbani followed by Sundargarh and Puri districts. As regards the production, it was highest for Phulbani followed by Ganjam and Sundargarh districts. He further observed that the variability in maize area in the state as well as across the districts increased over the years during the second period (1986-87 to 1998-99). The yield variability across the districts, as well as in the state level led to high variability in maize production.

Varghese and Rathore (2003) studied the maize production variability in Rajasthan and observed that the main contributing factor for inter-year production variability was due to yield variability of maize in all the selected districts. During pre and post green revolution periods, the variability in area was much less in comparison to yield and production. However, after 1986–87 the variability in area increased much as compared to production and yield.

Chand and Raju (2008) studied the instability in Andhra Pradesh agriculture and estimated instability in three major crops before 1980-81 to 1992-93 and after 1992-93 to 2003-04, the initiation of economic reforms at the state and district level in Andhra Pradesh and revealed that in a large state like Andhra Pradesh, the instability status as perceived through the state level data may be vastly different from that experienced at the disaggregate level. The study has concluded that the state level analysis does not reflect complete picture of shocks in agriculture production and farm income.

3.3 Review of Economics of Maize Production Studies

Sabur and Haque (1992) conducted a study on returns from some selected winter crops in Bangladesh and examined the profitability of four winter crops (potatoes, wheat, pulses and mustard) grown in Bangladesh from a sample of 200 farmers in Minishganj and Bogra district during the *rabi* season of 1987-88. Benefit-cost ratio was the highest for the Potato (2.29) followed by pulses, mustard and wheat. Potatoes were found to be the most profitable crop in spite of the fact that the production cost for potatoes were also highest of all the crops studied. Therefore, arrangements need to be made to provide credit, which will enable farmers to grow more potatoes.

Kumar (1995) compared the economics of different rice based competing crop sequences in Hoshiarpur district and found the expenditure incurred on insecticides and pesticides, irrigation, hired machinery, fuel and interest on running capital were comparatively higher in maize-wheat system. The study concludes that wheat in paddy-wheat crop sequence yielded higher net returns than wheat in maize-wheat crop system.

Sharma *et al.* (2003) carried out economic analysis of basmati vis-à-vis non-basmati rice based on data collected from the sample of 100 farm holdings selected from Jammu and Kathua districts of J&K state during the year 1999-2000 and observed that the returns from basmati rice were ₹14964.97 per hectare compared to ₹10208.48 from non-basmati rice after meeting both the production and marketing cost. The size-wise analysis showed that the highest returns from basmati were found to be on large farms (₹16162.15) and lowest on small farms (₹13959.90). In case of non-basmati rice also, the highest returns were found to be from large farms (₹11006.57) and the lowest on small farms *i.e.* ₹9305.40 per hectare. Thus, it was clear from the study that production of basmati rice was more profitable than non-basmati rice in spite of its high yield. The price factor was mainly responsible for high returns from basmati rice.

Koshta and Chandrakar (2005) in comparative analysis of economic efficiency of resources in production of rice under irrigated and rainfed situation of Chattisgarh, found that the cost on materials and labour inputs were comparatively more in the production of irrigated rice than rainfed rice while the profitability was higher in irrigated rice.

Singh (2005) worked out the costs and returns, input-output relationship and to studied the resource use efficiency of canal irrigated vis-à-vis tube well irrigated paddy in Faizabad district of U.P. and observed that average per ha cost on canal irrigated and tube well irrigated paddy was Rs 13,522 and Rs 15,067, respectively.

Khan *et al.* (2005) estimated net returns from main crops in district Malakand and observed that per acre cost of cultivation for rice, maize and wheat crop were ₹6095, ₹40773 and ₹4104, respectively with the exception of 12 per cent, 2 per cent and 1 per cent, respectively for rice, maize and wheat. The total gross returns of rice, maize and wheat were ₹10932, ₹7745 and ₹8613, respectively and the net returns were ₹4837, ₹3672 and ₹4509, respectively for rice, maize and wheat. They further observed that rice gives more net return than wheat and maize in the study area.

Mohiuddin *et al.* (2007) conducted study at the sadar upazila of Kishoreganj district and observed that on an average per hectare yield of maize was 4.70 tonnes and the average gross margin was Tk. 18047/ha on TVC basis and on cash cost basis it was Tk. 26887/ha. Net return per hectare on the basis of full cost, cash cost and variable cost were estimated at Tk. 11739, Tk. 26887 and Tk. 18047, respectively. It was found that the average returns to labour were Tk. 109.52/man-day on full cost basis, Tk. 291.80/man-day on cash cost basis and Tk. 135.06/man-day on variable cost basis, respectively. On full cost basis, benefit cost ratio (BCR) was 1.37. They further reported that on the basis of cash cost and variable cost, the BCR were 2.66 and 1.72, respectively for maize production.

Kachroo and Kachroo (2007) studied the economics of rice in 4 villages of R.S Pura block of Jammu region and revealed that the per hectare yield of basmati rice was higher in Kotli Shah Daula (24.10 qtl/ha) followed by

Khanchak (24.00 qtl/ha), Kotli Merdian (23.01 qtl/ha) and Kotli Arjun Singh (22.21 qtl/ha). The average yield and the cost of production were worked out to be 23.33 qtl/ha and ₹17833, respectively. They further observed that all the costs *i.e.* A_1, A_2, B_1, B_2, C_1 and C_2 were variable under different villages. The average value of cost computed in A_1 was lowest (₹8013) than A_2 (₹8523), B_1 (₹10169), B_2 (₹15678), C_1 (₹12325) and C_2 (₹17834) and the net return over each rupee invested in rice production were higher in case of cost A_1 (2.86) followed by A_2 (2.70), B_1 (2.26), C_1 (1.86), B_2 (1.48) and C_2 (1.29) and was more than one in all costs, which indicated that cultivation of basmati rice is economically profitable.

Hasan (2008) studied cost and economic returns of maize cultivation in the northern region of Bangladesh and observed that the average variable cost of maize cultivation was Tk. 22,836/ha, which was the highest at Dinajpur (Tk. 23,458/ha) and the lowest at Panchagarh (Tk. 22,006/ha). Major portion of variable cost was occupied by human labor (50.5 per cent) followed by fertilizers (15.7 per cent) and seed cost (15.5 per cent) in Dinajpur, while in Panchagarh those variable costs were 47 per cent, 15.7 per cent and 12.4 per cent for human labour, seed rate and fertilizer application, respectively. On an average, total fixed cost was Tk. 4414/ha, major portion of which was covered by land use cost (Tk. 3,500/ha). The fixed cost varied only for interest on operating capital which was Tk. 928/ha in Dinajpur and Tk. 871/ha for Panchagarh. He further observed that total production cost of maize was Tk.27240/ha, which was higher at Dinajpur (Tk. 27,886/ha) and lower at lower at Panchagarh (Tk. 26,377/ha) and average gross return was Tk. 60,981/ha, gross margin was Tk. 38,145/ha and net return was Tk. 33,741/ha. Among the two regions, gross return, gross margin and net returns were the higher at Dinajpur compared to Panchagarh.

Kudi and Abdulsalam (2008) analysed the costs and returns of Striga tolerant Maize variety in Southern Guinea Savanna of Nigeria. The costs and returns analysis indicated that labour and fertilizer inputs accounted for greater parts of the total variable costs incurred in both varieties. Labour cost of the improved Striga tolerant maize variety and farmers' varieties were 56.82 per cent and 58.81 per cent while fertilizer cost were 36.65 and 37.96 per cent, respectively.It was also found that cultivation of Striga tolerant maize variety was highly profitable as indicated by a gross margin of N 94, 479.21/ha compared to a gross margin of N 15, 683.73/ha for the farmers varieties.

Chapter 4

Economic Efficiency of Maize Production in Jammu Region of J&K State

4.1 Introduction

In India, maize is emerging as third most important crop after rice and wheat. Maize has its significance as a source of a large number of industrial products besides its uses as human food and animal feed. Diversified uses of maize for maize corn, starch industry, corn oil production, baby corns, popcorns, etc., and potential for exports has added to the demand of maize all over world. In India the maize potential districts in 23 states are Andhra Pradesh, Arunachal Pradesh, Assam, Bihar, Gujarat, Himachal Pradesh, Jammu and Kashmir, Karnataka, Madhya Pradesh, Maharashtra, Manipur, Meghalaya, Mizoram, Nagaland, Orissa, Punjab, Rajasthan, Sikkim, Tamil Nadu, Tripura, Uttar Pradesh and West Bengal. To meet the growing demand, per hectare yield of maize is estimated to rise to 2.36 tonnes as against 1.7 tonnes currently by the end of 2020. Maize does possess tremendous potential in terms of feed for dairy, poultry and piggery agro-industries. In order to increase the production and productivity of maize, the government adopted the new approach for area expansion for maize in view of serious competition from food and cereal crops. Maize crop is one of the important crop among the cereals and evaluating their economic efficiency, resource use efficiency, technical efficiency, growth and instability, technological adoption index, determinants of technical efficiency and problems faced by the farmers will help the maize growers of this study area to a greater extent as how to make the uses of resources efficiently so that they can increase the yield of maize crop for meeting the today's needs of this crop for various uses and make their cultivation more profitable besides this will act as a guideline for the planning of policy makers/scientists. Keeping in view the importance of the maize crop, and the facts described, a research study entitled, *"Economic efficiency of maize production in Jammu region of J&K state,"* has been undertaken.

4.2 Importance of the Study

The comparison of yield differences of maize crop between research farms (3.7 tonnes/ha) and farmers field in the study area (2.1 tonnes/ha) was realized to be (1.6 tonnes/ha) and between demonstration field (3.4 tonnes/ha) and farmers field (2.1 tonnes/ha) was found to be 1.3 tonnes/ ha. It thereby, indicate that the existing potential is yet to be realized. It is possible to exploit the potential yield by applying and rearranging the existing level of input use. For that matter, technical inefficiency may be attributed to be one of the factors responsible for low yield at farmer's field, because changes in productivity occur due to changes in technology. Thus, the identification of the factors responsible for enhancing maize productivity should receive considerable attention. Available evidences in the last few years revealed that technological package via efficient utilization of scarce resources which have alternative uses may accelerate the pace of maize production. It is therefore, necessary to quantify current levels of technical efficiency so as to estimate losses in production that could be attributed to inefficiencies due to differences in socio-economic characteristics and management practices. A detailed examination of the farm efficiency in terms of technical, allocative and economic efficiencies for increasing productivity in a resource poor state like Jammu and Kashmir is equally important. The technical efficiency refers to the ability of the farm to achieve the maximum possible output with available resources, whereas allocative efficiency refers to the ability to obtain optimal allocation of given resources. Thus one way of reducing the cost of production is to increase farm output by increasing technical efficiency. Economic efficiency is the combination of both the technical and allocative efficiencies. The measurement of economic efficiency is thus not complete without a study of technical efficiency and it is the frontier production function that enables the measurement of technical efficiency of farmers (Elsamma and George, 2002). Thus the identification of the factors responsible for enhancing maize productivity demands considerable attention. Apart from its manifold uses in terms of households and industries its production per unit input use per unit time and area needs to be increased. In this book, an attempt has been made for working out of resource use efficiency, cost and returns, cost benefit ratio may help biological scientists, policy makers and extension workers to increase its production as well as productivity due to which we examines various aspects of economic efficiency of maize production in Jammu region of Jammu and Kashmir state so that suitable policy for enhancing maize production and productivity can be formulated.

4.2.1 Objectives of the Study

1. To study the status of maize production
2. to workout the resource use efficiency of maize production
3. to examine the technical efficiency of maize production in Jammu region
4. to identify the factors influencing technical efficiency in maize production

4.3 Methodology

The case study "*Economic efficiency of maize production in Jammu region of J&K state,*" was carried out during 2008-09. The sampling structure, experimental procedures and techniques adopted during the course of investigation have been described in this chapter.

4.3.1 Climate and Rainfall

The state of Jammu and Kashmir is situated between 32°.15′ and 37°.05′ north latitude and 72°.35′ east and 83°.20′ east longitude. The state is the northern most part of India and looks like its crown and is girdled by Tibet to the east, China and Afghanistan to the north, to its west is

Pakistan. To its south lie the states of Punjab and Himachal Pradesh and is divided into three climatic divisions *viz.* outer plain and outer hills, middle mountains and Kashmir valley and inner Himalayas (Ladakh). The outer plains and outer hill includes the Kathua and Jammu districts, extend up to Shivalak hills in the north. In summer hot dry winds from plains of Punjab make it very hot, dust storms are common with occasional rain. Rainfall occurs from July to September. Average rainfall is 1143 mm and temperature shoots up to 46°C during July to September. During December to February temperature is between 13.5°C to 20°C, rainfall is 150 mm and in upper reaches there is snow. In the region of middle mountains and Kashmir valley winters are very cold. There is Snowfall during winter and sub- zero temperature. In valley, winters are long and summers are short but pleasant. Average rainfall is 732 to 854 mm. The maximum temperature hardly ever goes beyond 35°C. In region of inner Himalayas days are hot and there are no clouds in the sky. Winters are very cold. Temperature falls as low as minus 23°C. The average annual rainfall is 976 mm and humidity level is very low.

The two districts which were purposely selected for the present study was Udhampur and Doda districts of Jammu region. Uhampur district falls under sub-tropical zone which has hot and dry climate in summer and cold climate in winter and Doda district falls under temperate zone. The temperature of Udhampur district rises to 42°C in summer and dips to 1.5 °C in winters. Most of the rainfall takes place in July, August and September. The annual mean temperature of Doda district is 30°C with the average rainfall as 35 inches per annum.

4.3.2 Collection of Data

All the relevant data required for working out resource use efficiency, technical efficiency and factors affecting on technical efficiency were collected by survey method with the help of specially designed schedules for the purpose. Collection of data was done by the personal interview method of the maize growers. These schedules were pre-tested before using for actual data collection. For calculating compound growth rates and instability for area, production and yield of maize, secondary data was collected from the various published sources such as bulletins of the ministry of agriculture, Govt. of India, Directorate of Economics and

Statistics, Govt. of India, Directorate of Economics and Statistics, Jammu and Kashmir Govt.

Information of the Maize Grower and his/her Family

GENERAL INFORMATION

1. Name of household: **2. Village**

3. Block **4. Age:**

5. Mode of transportation **6. Distance from market**

7. Education: *Primary Middle Secondary Sr. Sec Graduate Post Graduate*

8. Status of farm workers:

Male			**Female**			**Children**		
No	*Age*	*Occupation*	*No*	*Age*	*Occupation*	*No*	*Age*	*Occupation*

9. Farm Income: ₹/annum **10. Non-farm Income: ₹/annum**

11.Total cultivated area (ha) Owned (ha): Leased in (ha):

12. Farm inventories:

a. Buildings: i. Cost ₹_____________ ii. Year of construction ____________________

b. Machinery:

a. Tractor with implements with year of purchase______________________________

b. Tube well with implements with year of installation___________________________

c. Pumping set with year of purchase _____________________________________

d. Otters such as jeep, tempo, motor cycle implements with year of purchase_________

c. Equipments: (for plant protection): Price and year of purchase

a. Sprayer:_____________________________

b. Duster:______________________________

Others:____________________________

13. Cost of cultivation of field crops

Item	*Crop-Quantity/No*	*Variety-Value (Rs)*	*Crop-Quantity/No*	*Variety Value (Rs)*	*Crop-Quantity/No*	*Variety Value (Rs)*
Area (ha)						
Soil testing charges (Rs)						
Land preparation						
Ploughing-machine (hours)						
Machine labour (hours) except ploughing						
Own human labour (days)						
Hired human labour(days)						
No. of irrigation						
Seed (kg)						
FYM/compost(q)						
Fertilizers(Kg)						
N						
P						
K						
Plant protection						
Intercultural						
Weeding						
Harvesting						
Threshing						
Others(specify)						
Production(q)						

Rental value of land (RS/ha):____________________

Price of main Product (Rs/q):____________________

Price of by-product (Rs/q):____________________

14. Management Practices:

a) Methods of Ploughing:

i. Tractor: Yes/No
ii. Bullock : Y/N
iii. Both : Y/N
iv. Rates:
v. Tractor : ₹/Hours
vi. Bullock: ₹/Day
vii. Women: ₹/Day
viii. Men : ₹/Day
ix. Children : ₹/Day

b) Seed:

i. Changing seed varieties: every yr./2nd yr./3rd yr.
ii. Spacing of plant as per recommendation: Y/N
iii. Sowing timely: Y/N
iv. Seed testing for germination: Y/N
v. Seed treatment for disease: Y/N (HYV/Others)
vi. Source of Seeds : Owned/exchange with other Farmers/Purchased
vii. Seed Purchased : Neighbor/Private/NSC/Govt./Uni./any other

c) Method and Time of seed sowing:

i. Broadcasting
ii. Lining
iii. Zero tillage
iv. Normal Y/N
v. Early Y/N
vi. Late Y/N
vii. Off-season : Y/N

d) Organic/inorganic nutrients use and method of application :

Use of FYM : Qty________ ₹__________

Compost: Qty________ ₹__________

Vermin-compost: Qty_________ ₹_________________

Green-Manure : Y/N

Use of Fertilizers :

As per recommendation to specific crop.: Y/N

Knowledge for fertilizer adulteration testing : Y/N

Use of Micronutrients:

Zink :	Kg	Rs
Sulphur:	Kg	Rs.
Iron :	Kg	Rs.
Others :	Kg	Rs.

e) Pesticides:

Pesticide purchased Private/Govt./Others

Test of pesticide for quality: Y/N

Pesticides: Chemical/Bio-pesticides with name:

f) Method of irrigation: Flooding/Sprinkler/Drip/Other

g) Method of weeding: By hand/By machine/Cultural/Chemicals/Other

15. Constraints in crop production:

a) Production

i. High inputs cost: Y/N
ii. High labour cost: Y/N
iii. Availability of labour : Easily/difficult
iv. Out migration of labour : Y/N
v. Weeds : Y/N
vi. Very small holdings : Y/N
vii. Non availability of disease – resistance varieties : Y/N
viii. Humidity/fog: Y/N
ix. High cost of pesticides : Y/N
x. High Cost of water for irrigation : Y/N
xi. Decline in soil fertility :Y/N
xii. Nematodes/White lice : Y/N
xiii. High Cost of micronutrients : Y/N
xiv. Availability of micronutrients: In time/not in time
xv. Quantity of micronutrients: adequate/inadequate

b) Marketing

i. Markets: accessible/inaccessible
ii. Distance of market : Far/Very far/moderate
iii. High cost of transportation : Y/N
iv. Transport facilities : adequate/inadequate
v. High cost of storage: Y/N
vi. High Loss due to spoilage during transportation, storage etc.
vii. Agricultural income is : Profitable/non-profitable

c) Institutional

i. Credit availability: in time/not in time
ii. Quantity of credit : adequate/inadequate
iii. Rate of interest: High/moderate
iv. Banks: Accessible/inaccessible
v. Absence of scientific/technical knowledge
vi. Lack of extension workers: Y/N
vii. Visits of extension workers: weekly/fortnightly/monthly/bi-monthly/ six monthly/no visit.
viii. Future of agriculture : very good/good/moderate/poor: Y/N
ix. Poor infrastructure of the area
No school
Number of Banks : Less
Distance of banks : Far/Very far
Medical facilities : no dispensary/no hospital/no doctor etc.

4.3.3 Sampling Structure

A multi stage sampling was adopted for the selection of samples. Districts, blocks, villages and farmers were the first, second, third and fourth stage units, respectively. Udhampur and Doda districts of Jammu region were selected purposively because these two areas covered the maximum area under maize cultivation (Doda covered 23.47 per cent and Udhampur covered 25.35 per cent out of the total area under maize cultivation in Jammu region). Further, out of the total maize production of Jammu region, Udhampur district had 31.80 per cent and Doda district had 30.08 per cent production. Three blocks from each district were selected randomly as secondary stage units. From each block two villages were selected as the third stage units and the ultimate units were twenty households from each village so as to constitute sample size of 240 households. For each selected villages, the list of maize cultivators was prepared by making into four size group of holdings as mentioned below:

(i) Marginal: Land up to 1 hectare
(ii) Small: 1.01 hectare to 2 hectare
(iii) Medium: 2.01 hectare to 3 hectare
(iv) Large: land above 3.01 hectare

The cultivators from all the villages were selected randomly on the basis of holding size.

4.3.4 List of Selected Blocks

Following six blocks falling in Doda and Udhampur district were selected for this study:

District Doda

i) Bhaderwah
ii) Doda
iii) Assar

District Udhampur

i) Udhampur
ii) Duddu
iii) Ramnagar

4.3.5 Selection of Villages

The details of the selected villages are as under

Blocks Selected	*Villages Selected*
(i) Bhaderwah	a) Neyota
	b) Marela
(ii) Doda	a) Brotha
	b) Boghata
(iii) Assar	a) Kothyara
	b) Jathi

Blocks Selected	Villages Selected
(iv) Udhampur	a) Tikkri
	b) Manda
(v) Duddu	a) Majouri
	b) Rasli
(vi) Ramnagar	a) Thaplal
	b) Gagote

4.3.6 Analysis of Data

Some of the important points in the analysis of the data and the methodology adopted has been elucidated below:

4.3.7 Cost Concepts

The various cost concepts were used for the study and employed to explain cost and returns structure of the farm.

Cost A_1: Expenditure on casual labour, bullock labour, farm machinery, seeds fertilizer and manure, plant protection chemicals, irrigation, miscellaneous expenditure (cost of transportation, baskets etc) and interest on working capital + depreciation + land revenue.

Cost A_2: Cost A_1 + rent paid for leased -in land.

Cost B_1: Cost A_1 + interest on value of owned fixed capital excluding land.

Cost B_2: Cost B_1 + rental value of owned land (net of land revenue) + rent paid for leased-in land.

Cost C_1: Cost B_1 + imputed value of family labour.

Cost C_2: Cost B_2 + imputed value of family labour.

4.4 Quantification of Variables

4.4.1 Seed

This item included the quantity of seed used per hectare in kilograms at the farm for the maize crop under study. Value of the seed produced at the farm was assessed at the prevailing market price to workout its cost whereas the cost calculation of the seed, which was purchased from the market, was based on the actual payments made for its purchase.

4.4.2 Fertilizer + Manure

This included the expenditure incurred on purchase of chemical fertilizers + farmyard manures used for the production of crop on the sample farms. The farmyard manure used at the farm was assessed at the prices prevailing in the study area. Similarly, the physical quantities of different fertilizer used were multiplied with the market price.

4.4.3 Bullock Labour

Bullock was used on the farm to accomplish major agricultural operations, like preparatory tillage, sowing, transporting input and output from one farm to the other farm etc. All these operations were carried on either by owned bullock or by arranging it on exchange or hire basis. Bullock hours actually used to perform any of these farm operations per hectare of crop raised on the sample farms were recorded to measure the amount of animal power used. However, total bullock hour were taken into consideration for functional analysis of the variables.

4.4.4 Human Labour

The human labour was employed to accomplish for various farm operations on the sample farms. It comprised family labour, attached farm servant and hired casual labour. Existing casual labour wage for different operations as used to work out the total wage bill of labour employed per hectare of maize crop.

4.4.5 Irrigation Charges

The main source of irrigation in the study area of district Udhampur and Doda (irrigated) was canal water. The information on irrigation was thus obtained in terms of irrigation hours put in to irrigate maize crop with canal water.

4.4.6 Variable Costs

It included expenditure made on various input components *viz.* seeds, chemical fertilizers + manures, plant protection, irrigation, human labour, diesel oil and machinery used per hectare for any crop and interest on working capital.

4.4.7 Fixed Costs

It included expenditure made on various fixed components *viz.* rental value of owned land, depreciation on implements and farm building, rent paid for leased in land and interest on fixed capital.

4.4.8 Gross Returns

Gross returns from maize crop enterprise per hectare were obtained by the value of main product and that of the by-product realized per hectare from the crop. Prices of main product and the by-product used to arrive at gross returns were the post-harvest market prices in the study area.

4.5 Status of Maize

4.5.1 Estimation of Growth and Instability

In order to study the status of maize, the present study made use of time-series data on different variables like area, production and yield of maize for India, Jammu and Kashmir State, Jammu region and for selected districts separately. For working out the growth rates of area, production and yield of maize, the time period had been divided into three different periods as:

(i) Period I (1987-88 to 1996-97)

(ii) Period II (1997-98 to 2006-07) and

(iii) Overall Period (1987-88 to2006-07)

In order to work out the trend of variables like area, production and yield, compound growth rate as used by Kachroo and Kachroo, 2006 calculated as follows:

$$y\text{-}_t = y_0\,(1+g)^t \text{ or } a\,(1+g)^t$$

$$yt = ab^t \text{ (Where } b = 1 + g)$$

where,

y: Absolute value

a: initial value of y or constant

t: time (years)

b: Estimated regression coefficient (parameter)

On log transformation

$$\log y = \log a + t \log b$$

Then, compound growth rate (r) = (b-1)* 100 was estimated with the help of computing package SPSS, 7.5. version.

The growth rates were tested statistically for their significance through t- test as given below :

$$t = r/S.E.(r) \sim t\,\alpha_{,n-2}$$

$$S.E. = b\sqrt{[\{\Sigma (\log y)^2 - (\Sigma \log y/n) - \log b^2\}^2\ \Sigma xi^2]/0.43429(n-\alpha)\ (\Sigma x_i^2)x100}$$

Period wise instability of these variables was measured through coefficient of variation which was calculated as:

$$\text{Coefficient of variation (C.V.)} = \frac{\sigma}{X} \times 100$$

4.5.2 Decomposition of Growth Analysis

For analyzing effect of area, productivity and their interaction in increasing the maize production was examined by using differential equation as used by Kachroo and Sharma (2008).

$$\Delta P = Y_0\,\Delta A + A_0\,\Delta Y + \Delta A\,\Delta Y$$

where,

$\Delta A = A_n - A_{0,}\ \Delta Y = Y_n - Y_{0,}\ \Delta P = P_n - P_0$ and $A_{0,}\ Y_{0,}\ P_0$ are the area, productivity and production, respectively in base year, whereas A_n, Y_n and P_n are area, productivity and production in current year, respectively and ΔA, ΔY, ΔP are the changes in area, productivity and production, respectively. Contribution of area, productivity and their interaction to total production differential of maize was obtained separately for each of the period under study as also for total period of study (1987-88 to 2006-07) on all India, state, region and selected district basis.

The data comprising of 20 years was divided into two phases each with 10 years *i.e.*

(i) Period I (1987-88 to 1996-97)

(ii) Period II (1997-98 to 2006-07)

Thus, the changes in production (ΔP) were due to:

(i) Area effect ($Y_{0.} \Delta A$)

(ii) Yield effect ($A_{0.} \Delta Y$)

(iii) Interaction of area and yield effect ($\Delta A_{.} \Delta Y$)

4.5.3 Technological Adoption Index

The technological adoption index (TAI) is a measure of technology adoption practices by the farmers which including area under high yielding varieties, appropriateness of irrigation level and dosages of fertilizers (Anupama *et al.*, 2005). The TAI of farmers was computed by using following formula;

$$TAI_i = \frac{1}{5}\left[\frac{AH_i}{CA_i} + \frac{NA_i}{NR_i} + \frac{PA_i}{PR_i} + \frac{KA_i}{KR_i} + \frac{IA_i}{IR_i}\right] \times 100$$

where,

I: Numbers of farmers, say 1,2, 3,...,n

TAI_i: Technology adoption index of the ith farmer

AH_i: Area under modern maize varieties (ha)

CA_i: Total area of maize (ha)

NA_i: Quantity of nitrogen applied for maize (kg/ha)

NR_i: Recommended dose of nitrogen for maize crop (kg/ha)

PA_i: Quantity of phosphorus applied for maize (kg/ha)

PR_i: Recommended dose of phosphorus for maize crop (kg/ha)

KA_i: Actual amount of potash applied for maize (kg/ha)

KR_i: Recommended amount of potash applied for maize (kg/ha)

IA_i: Actual number of irrigation applied

IR_i: Recommended number of irrigation for maize crop.

TAI ranges between 0 to 100.

The index was calculated for all the 240 maize growers. Thereafter, they were classified into three categories *viz.*, low adopters (< 33 per cent TAI), medium adopters (33- 66 per cent TAI) and high adopters (> 66 per cent TAI).

4.6 Measurement of Resource Use Efficiency

It is a key factor for increasing productivity. Technical and allocative efficiencies were employed to measure the resource use efficiency. The technical efficiency in production was estimated by using the stochastic frontier production function. The regression co-efficient of factor input from Cobb-Douglas production function (OLS) were used to calculate the Marginal Value Production (MVP) at Geometric mean level for the average farms. In order to study resource – allocative efficiency, the ratio of MVP of a respective input to the marginal factor cost (MFC) for each input was compared and tested for its equality to 1, *i.e.* MVP/MFC= 1 (Yotopoulos, 1967).

If the ratio is equal to one, it means the input is used optimally. If the ratio is more than one, the input is under-utilized and if the ratio is less than one, the input is over-utilized. To calculate Marginal Value Productivity (MVP) of resource x_i, the following formula had been used.

$$MVP = \frac{b_i\,[GM(Y)}{GM\,(X_i)]}$$

where,

MVP (x_i) is the Marginal Value Productivity of ith resource:

b_i is the regression coefficient

GM (Y) is the Geometric Mean of Output

GM (X_i) is the Geometric Mean of Inputs

4.7 Specification of the Model

The stochastic frontier production function of the Cobb-Douglas type was specified for the present study and estimated by using Limdep computing package. The Cobb-Douglas Functional form was used because, the functional form has been widely used in farm efficiency measurement for the developing and developed countries, the functional form meets the requirement of being self-dual allowing an examination of economic efficiency and lastly Kopp and Smith (1980) suggested that functional form has a limited effects on empirical efficiency measurement. Due to its advantage over the other functional forms, it is widely used for the estimation of technical efficiency as well as resource use efficiency in the frontier production studies. This model was used by many researchers for developing their studies few of them were Hazarika and Subramanian (1999), Anupama *et al.* (2005), Aldeleke *et al.* (2008), Ogundari and Ojo (2007) and Dolisca and Jolly (2008).

The model used as:

$$\text{In } y_i = \beta_{i0} + \beta_{i1} \text{ In } L + \beta_{i2} \text{ In } F + \beta_{i3} \text{ In } K + \beta_{i4} \text{ In } I + v_i - u_{i\,(i = 1, 2, \ldots, n)}$$

where,

Y_i: Yield of maize in the ith farm (q/ha)

L: Human labour used in maize crop (mandays/ha)

F: Quantity of fertilizer (N+P+K) used (kg/ha) in maize crop

K: Capital which included overhead expenditure on animal and machine labour and seeds (Rs/ha)

I: Irrigation no. of times applied

v_i-u_i: Random error-term

The number of variables to be included in the model is generally determined by the nature of economic phenomenon under investigation and the purpose of research care should be taken in including only the most important variables in the model. Whenever compromises are necessary, care should be taken to omit such variables as are relatively least important.

4.8 Determinants of Technical Efficiency

It was used to identify the socio-economic factors influencing the technical efficiency at the farm level. MLE (Maximum likelihood estimators) estimates of technical efficiency, regressed on rental value of per gross cropped area (Proxy for land quality), proportion of females in total agricultural workers in the family, proportion of children in the family, education dummy for the household having family adult member with education above primary level and farm size. As the technical efficiency variable varies between 0 and 1, the variable will transformed into [TE/(1- TE)] so that the later transformed variable now varies between – ∞ and + ∞, which will facilitate estimation of the parameters by using the OLS technique.

The following linear regression model was used to identify the socio-economic factors that condition technical efficiency of sample farms.

$$L_n\ [TE/(1- TE)] = \beta_0 + \beta_1 X_{1ij} + \beta_2 X_{2ij} + \beta_3 X_{3ij} + \beta_4 X_{4ij} + \beta_5 X_{5ij} + u_i$$

where,

TE_{ij}: Technical efficiency for ith crop on j-th farm,

β_0: Intercept/Constant

β_i: Regression coefficients,

X_1: Age of the head of family

X_2: Proportion of female workers in total agricultural workers in the family

X_3: Proportion of children in the family as helper

X_4: Dummy for adult members/having education above primary level

X_5: Farm size and

u_i: Error term.

4.9 Results

The results of the present study are given in this chapter. Maize is emerging as an important cereal crop in India after rice and wheat. Though, Jammu and Kashmir State is one of the traditional maize growing states in India, the productivity of maize is very low when compared to the national average thereby falling under the category of low producing states. Therefore, efforts need to be taken to enhance productivity of maize in the state. The present study was undertaken with the objectives of examining the performance of maize crop to find the adoption level of modern maize technology and its economic impact, resource use efficiency and also estimate the technical efficiency of maize growers and factors effecting on technical efficiency.

4.9.1 Production Status of Maize

Estimation of compound growth rates, instability and decomposition of maize crop production are included in this section. Compound growth rates and instability for area, production and yield of maize were worked out to know the status of maize in India, Jammu and Kashmir State, Jammu region and selected districts.

4.9.2 Growth Performance of Maize Crop

The perusal of the data of last 20 years (Table 4.1 and Figure 4.1) indicated that during first period (1987-88 to1996-97) area, production and yield experienced positive growth rates with their values at 0.859 per cent, 3.637 per cent and 3.531 per cent, respectively at 1 per cent level of significance.

Table 4.1: Growth Trend of Area, Production and Yield of Maize Crop in India

S.No.	*Particulars*	*Period I (1987-88 to 1996-97)*	*Period II (1997-98 to 2006-07)*	*Overall Period (1987-88 to 2006-07)*
1.	Area	0.859* (0.002)	2.658* (0.002)	0.870* (0.001)
2.	Production	3.637* (0.015)	3.806* (0.007)	3. 132* (0.002)
3.	Yield	3.531* (0.012)	0.838* (0.006)	2.299* (0.001)

* 1 per cent level of significance. Figures in the parentheses denote standard errors of their respective coefficients.

The data further indicated that during period second (1997-98 to 2006-07), the trend for production was somewhat same as period I *i.e.*, 3.806 per cent whereas for area and yield it had totally changed. During period II, yield trend was less than one per cent (0.838 per cent) and area trend was more than 2 per cent (2.658 per cent) and they are significant at 1 per cent level of probability.The data further, showed that maize crop registered a significant growth of 0.870 per cent in area, 3.132 per cent in production and 2.299 per cent in yield, respectively for the overall period of twenty years (1987-88 to 2006-07).

The data in Table 4.2 and Figure 4.2 (a) revealed that during the period first (1987-88 to 1996-97), growth rates for area (0.441), production (3.620) and yield (3.191) were positive and significant at 1 per cent level of probability in Jammu and Kashmir state. The growth rate in production and yield of maize during the second period (1997-98 to 2006-07) though significant at 1 per cent level of probability, declined by (-) 0.070 per cent and (-) 0.419 per cent, respectively. But as far as area was concerned, it had a positive growth trend of 0.350 per cent. During the overall period

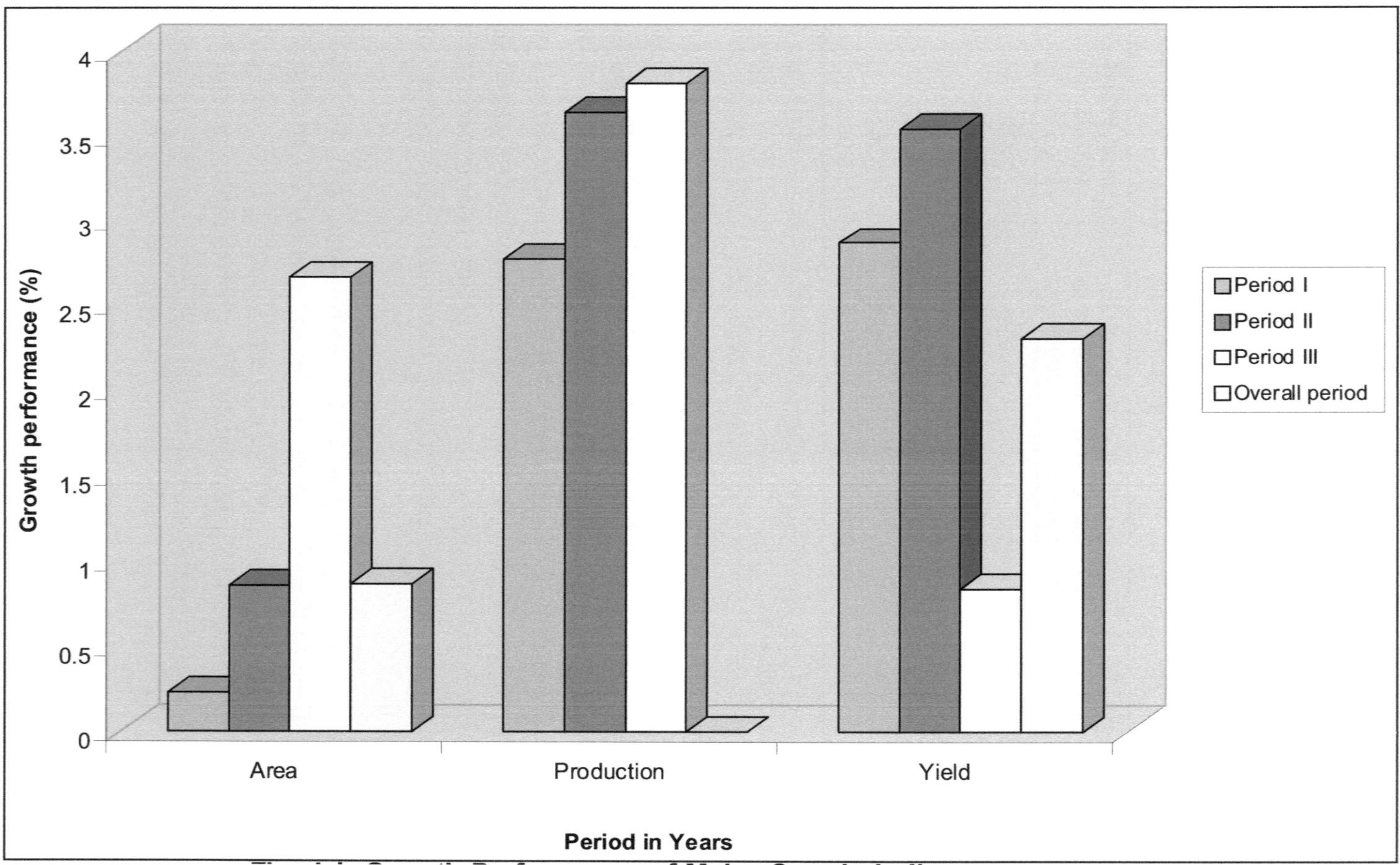

Figure 4.1: Growth Performance of Maize Crop in India.

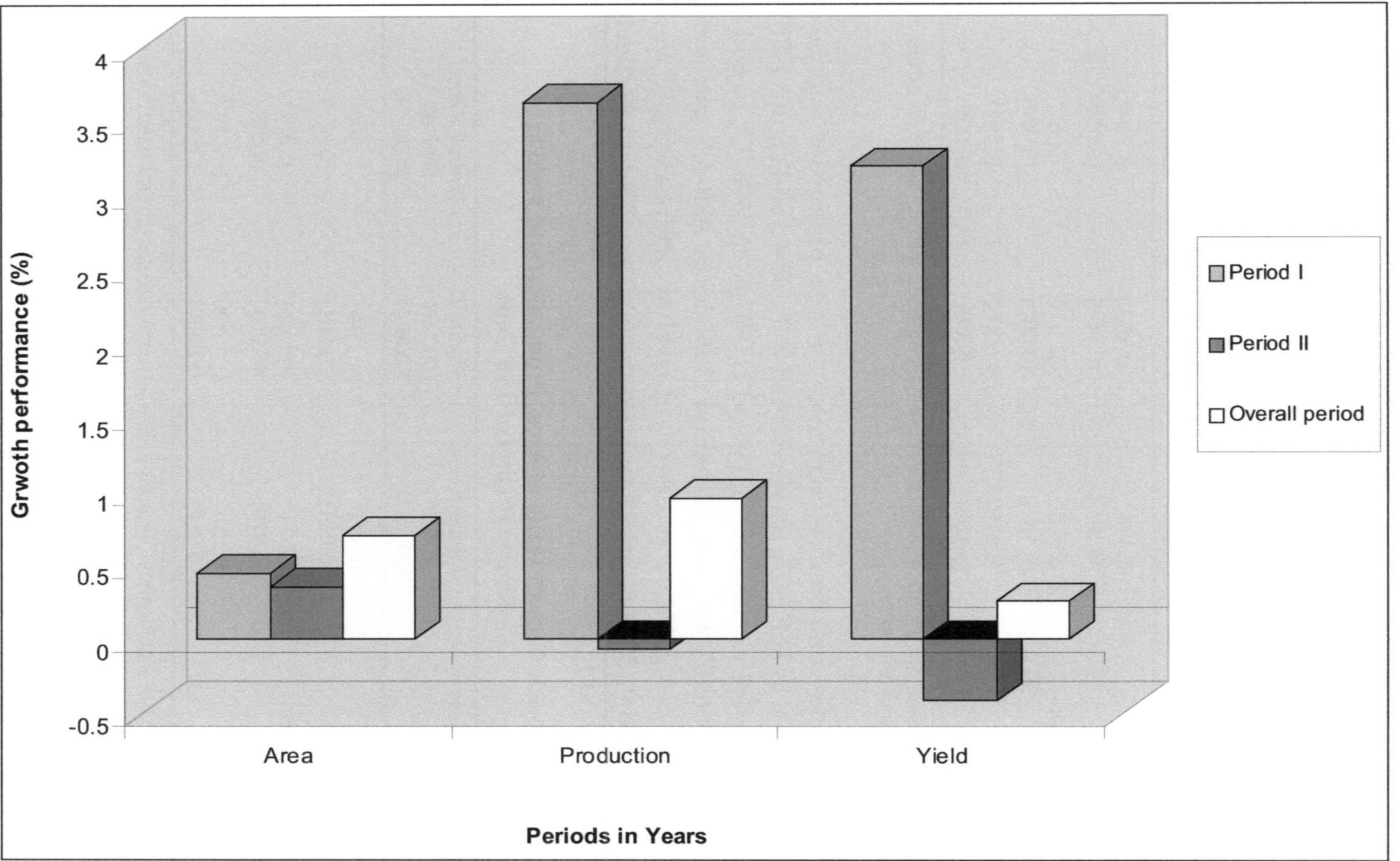

Figure 4.2(a): Growth Performance of Maize Crop in J&K State.

(1987-88 to 2006-07), area, production and yield trend was 0.701 per cent, 0.952 per cent and 0.252 per cent, respectively.

Table 4.2: Growth Trend of Area, Production and Yield of Maize Crop in Jammu and Kashmir State

S.No.	*Particulars*	*Period I (1987-88 to 1996-97)*	*Period II (1997-98 to 2006-07)*	*Overall Period (1987-88 to 2006-07)*
1.	Area	0.441* (0.001)	0.350* (0.002)	0.701* (0.008)
2.	Production	3.620* (0.161)	(-)0.070* (0.090)	0.952* (0.004)
3.	Yield	3.191* (0.162)	(-)0.419* (0.09)	0.252* (0.005)

* 1 per cent level of significance. Figures in the parentheses denote standard errors of their respective coefficients.

Similarly, the compound growth rates for area, production and yield in Jammu region of J&K State was presented in Table 4.3 and Figure 4.2(b). Growth rates were worked out for period I (1987-88 to 1996-97), period II (1997-98 to 2006-07) and overall period (1987-88 to 2006-07) and were found positively significant at 1 per cent level of significance and were 0.633, 2.511 and 0.155 per cent for area, production and yield, respectively during the period I.

Table 4.3: Growth Trend of Area, Production and Yield in Jammu Region of J&K State

S.No.	*Particulars*	*Period I (1987-88 to 1996-97)*	*Period II (1997-98 to 2006-07)*	*Overall Period (1987-88 to 2006-07)*
1.	Area	0.633* (0.001)	0.269* (0.001)	0.800* (0.007)
2.	Production	2.511* (0.010)	(-)1.224* (0.001)	0.351* (0.003)
3.	Yield	0.155* (0.012)	(-)1.496* (0.009)	(-)0.583* (0.004)

* 1 per cent level of significance. Figures in the parentheses denote standard errors of their respective coefficients.

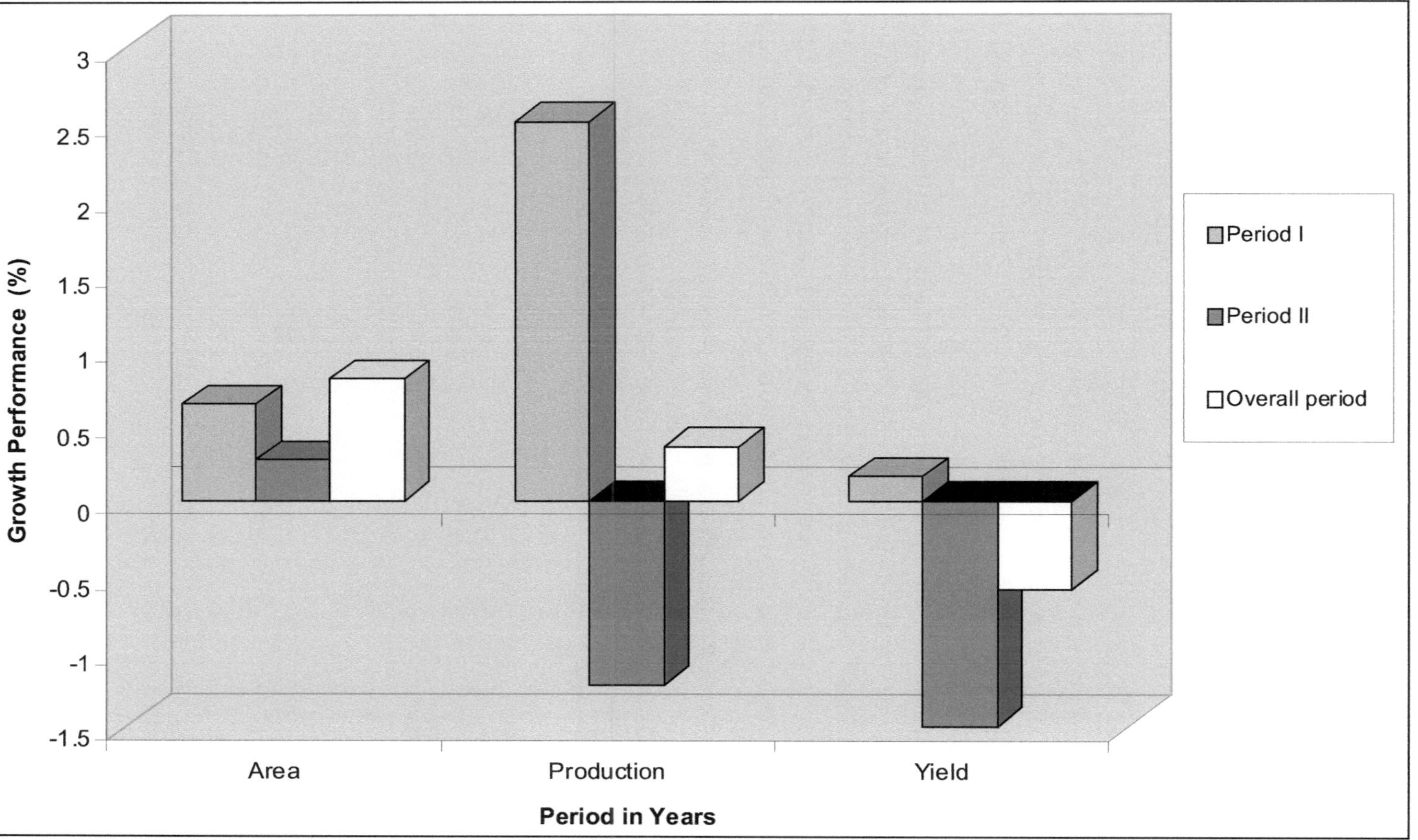

Figure 4.2(b): Growth Performance of Maize Crop in Jammu Region.

The Table 4.3 further indicated that period II (1997-98 to 2006-07) showed negative growth trend for production (-1.224 per cent) and yield (-1.496 per cent) while as for area, growth trend was positively significant with 0.269 per cent but the increase was less than the previous period.

During the overall period (1987-88 to 2006-07), the growth rate of yield decreased by 0.583 per cent but area (0.800 per cent) and production (0.351 per cent) of maize experienced positive and significant growth rates across the region.

Compound growth rates for area, production and yield of maize were also worked out for major maize producing districts of Jammu region over a period of 20 years from 1987-88 to 2006-07 and is presented in Table 4.4 and Figure 4.2 (c). District wise growth rates indicated that there was considerable variation in the performance of different districts of Jammu region. Jammu district recorded the highest growth rate of 1.469 per cent and 2.047 per cent in area and production, respectively, and growth trend for yield (0.792 per cent) was less than Kathua district (1.558 per cent). In case of area, it was significantly positive with less than 1 per

Table 4.4: Growth Trend of Area, Production and Yield in Major Maize Growing Districts of Jammu Region from 1987-88 to 2006-07

Sl.No.	*Districts*	*Area*	*Production*	*Yield*
1.	Jammu	1.469* (0.006)	2.047* (0.006)	0.792* (0.002)
2.	Kathua	0.275* (0.004)	0.887* (0.008)	1.558* (0.004)
3.	Udhampur	1.371* (0.022)	-0.422* (0.005)	-1.462* (0.005)
4.	Doda	0.685* (0.001)	-3.637* (0.099)	-4.021* (0.019)
5.	Rajouri	0.630* (0.001)	1.032* (0.005)	0.493* (0.005)
6.	Poonch	0.363* (0.001)	-0.727* (0.005)	-0.606* (0.005)
7.	Jammu Region (maize growing districts)	0.800* (0.001)	0.351* (0.003)	-0.583* (0.003)

* 1 per cent level of significance. Figures in the parentheses denote standard errors of their respective coefficients.

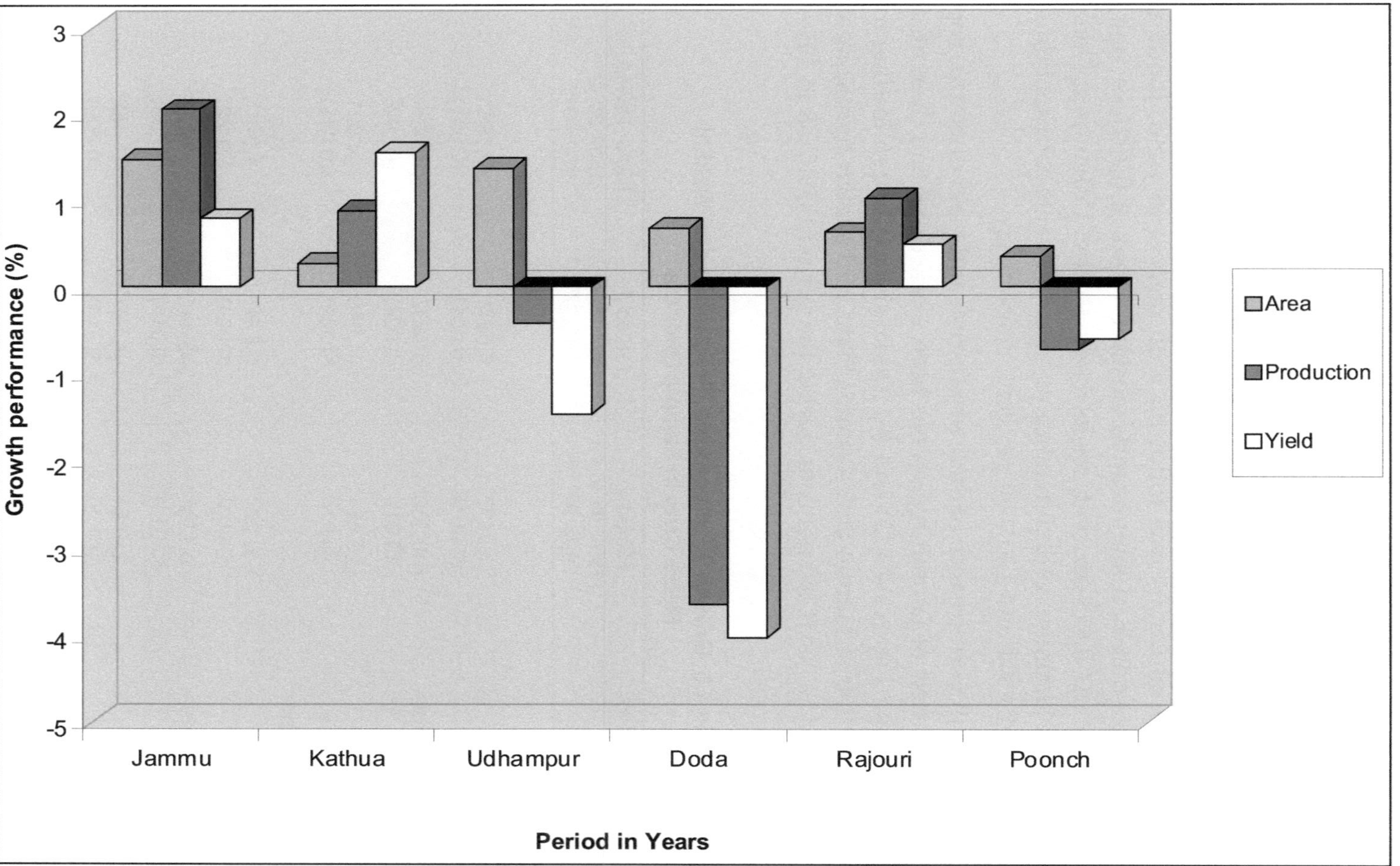

Figure 4.2(c): Performance of Major Maize Growing Districts of Jammu Region.

cent growth rate barring two districts of Jammu and Udhampur where it was more than 1 per cent. With regard to production, 50 per cent of the maize growing districts were showing positively significant trend (*i.e.*, Jammu 2.047 per cent, Kathua 0.887 per cent, and Rajouri 1.032 per cent) and 50 per cent were showing significantly negative trend *i.e.*, Udhampur (-0.422 per cent), Doda (-3.637 per cent) and Poonch (-0.727 per cent). Same trend (production) was followed with respect to yield. Jammu, Kathua and Rajouri had significantly positive trend of 0.792 per cent, 1.558 per cent and 0.493 per cent, respectively. District Udhampur, Doda and Poonch of Jammu recorded significantly negative value corresponding to (-) 1.462, (-) 4.021 and (-) 0.606 per cent, respectively.

The growth performance of area, production and yield of maize crop in sample districts of Jammu region has been worked out and presented in Table 4.5 and Figures 4.3 (a) and (b). Doda district though leading in maize production of Jammu and Kashmir State, had shown significantly negative trend *i.e.*, (-) 0.360 per cent during the period I (1987-88 to 1996-

Table 4.5: Growth Trend of Area, Production and Yield of Maize Crop in Sample Districts

S.No.	*Particulars*	*Period I (1987-88 to 1996-97)*	*Period II (1997-98 to 2006-07)*	*Overall Period (1987-88 to 2006-07)*
	Doda			
1.	Area	1.247* (0.001)	0.587* (0.001)	0.685* (0.001)
2.	Production	(-)0.360* (0.01)	(-)1.007* (0.038)	(-)3.637* (0.099)
3.	Yield	(-)0.684* (0.01)	(-)1.586* (0.038)	(-)4.021* (0.019)
	Udhampur			
1.	Area	(-)0.204* (0.004)	0.674* (0.004)	1.371* (0.022)
2.	Production	0.179* (0.011)	(-)3.441* (0.016)	(-)0.422* (0.005)
3.	Yield	1.435* (0.008)	(-)4.087* (0.014)	(-)1.462* (0.005)

* 1 per cent level of significance. Figures in the parentheses denote standard errors of their respective coefficients.

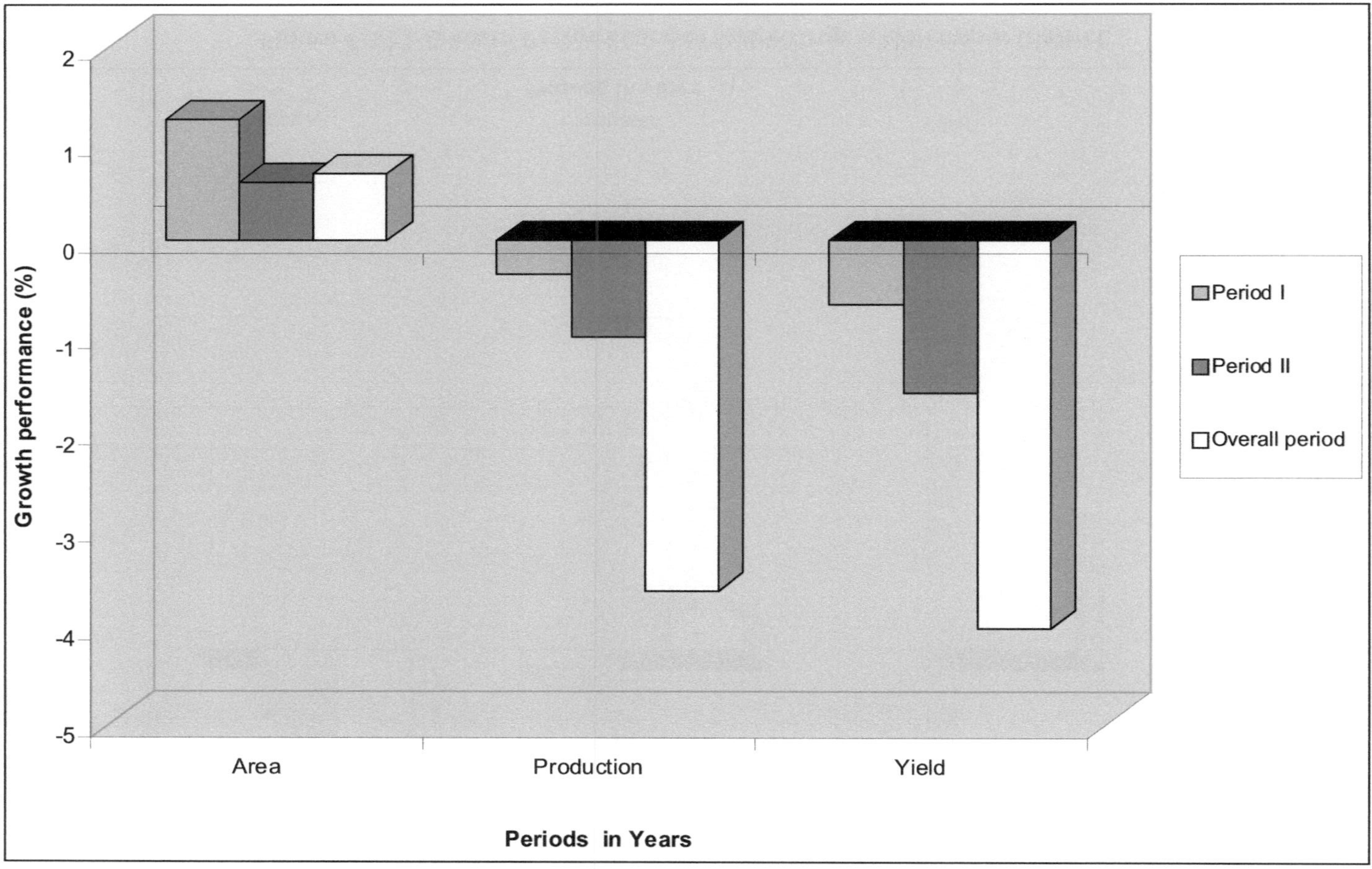

Figure 4.3(a): Growth Performance of Maize Crop in Doda District.

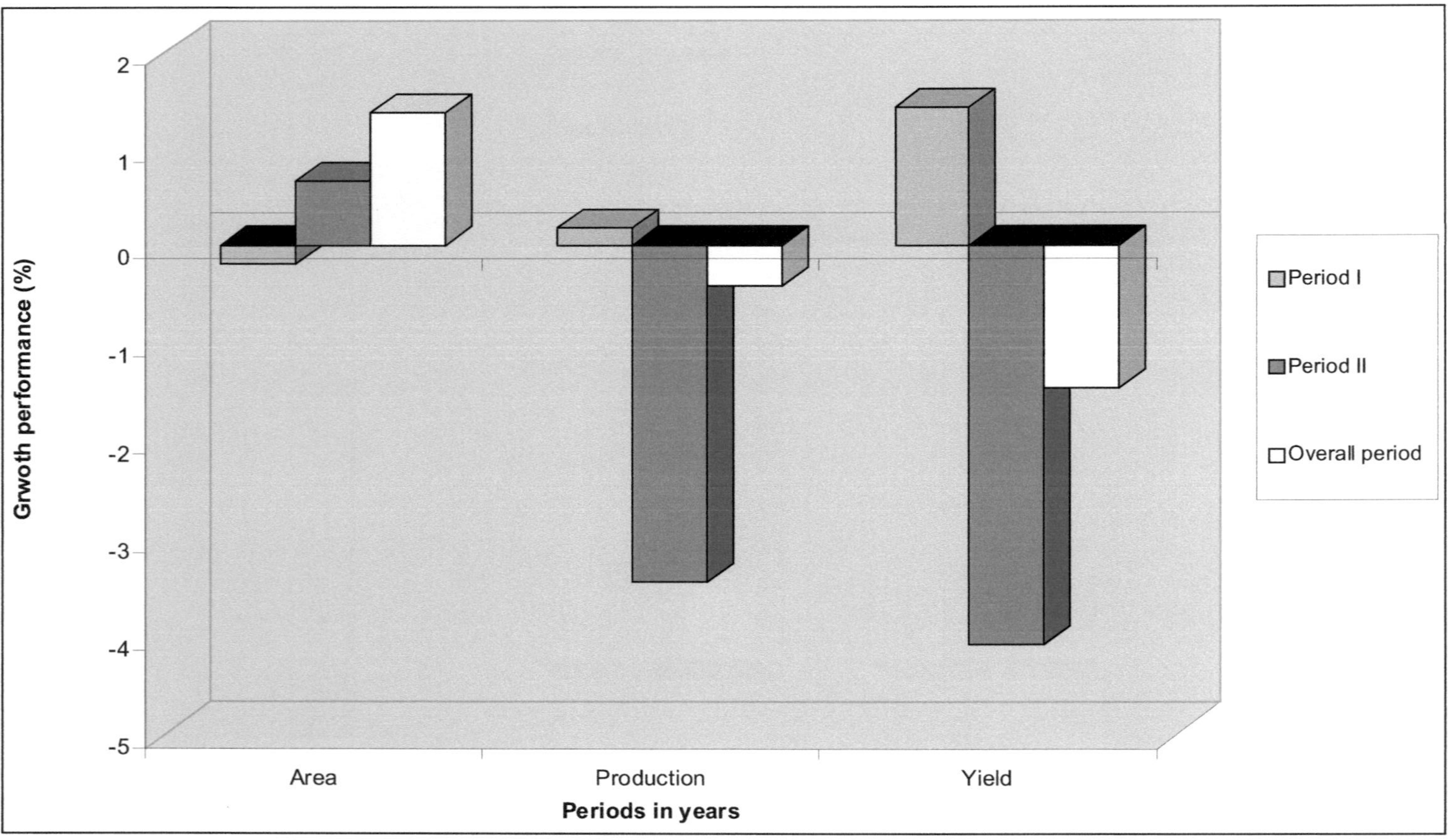

Figure 4.3(b): Growth Performance of Maize Crop in Udhampur District.

97) for production and for yield (-) 0.684 per cent. Period II (1997-98 to 2006-07) followed the similar trend with (-) 1.007 per cent for production and (-) 1.586 per cent for yield. Area was the only variable with significant positive trend for both periods *i.e.,* 1.247 per cent and 0.587 per cent, respectively. However, for the overall period (1987-88 to 2006-07), the growth trend for area, production and yield were 0.685 per cent, (-) 3.637 per cent and (-) 4.021 per cent, respectively.

The Table 4.5 also presented growth performance of maize in Udhampur district. The growth rate for area (-0.204 per cent) was negative but significant in period I (1987-88 to 1996-97) and positive for production and yield *i.e.,* 0.179 per cent and 1.435 per cent, respectively.

During the period II (1997-98 to 2006-07) the growth rate for area was positive *i.e.* 0.674 per cent and negative for production (-3.441 per cent) and yield (-4.087 per cent), but significant at 1 per cent level of probability. For the overall period (1987-88 to 2006-07), there was negative and significant growth rates for production (-0.422 per cent) and yield (-1.462 per cent), while positive and significant for area (1.371 per cent).

4.9.3 Instability in Maize Crop Cultivation

The instability of maize crop in India (Table 4.6) with respect to area, production and yield indicated that during period I, area was having minimum instability of 3.072 per cent and production with maximum instability of 14.893 per cent followed by yield instability of 13.382 per cent. But during period II, yield with minimum instability (6.480 per cent,) and production with maximum instability 13.350 per cent followed by area 8.401 per cent. During overall, the period the coefficient of variation for area, production and yield was 9.551 per cent, 29.340 per cent and 20.910 per cent, respectively.

Table 4.6: Instability of Area, Production and Yield of Maize Crop in India

S.No.	*Particulars*	*Period I (1987-88 to 1996-97)*	*Period II (1997-98 to 2006-07)*	*Overall Period (1987-88 to 2006-07)*
1.	Area	3.072	8.401	9.551
2.	Production	14.893	13.350	29.340
3.	Yield	13.382	6.480	20.910

The instability in area, production and yield of maize crop in Jammu and Kashmir State is presented in Table 4.7. During the period I, the instability in area, production and yield was 1.850 per cent, 15.180 per cent and 14.612 per cent, respectively. However, the instability in period II was 2.071 per cent, 7.322 per cent and 11.090 per cent for area, production and yield, respectively. The overall instability was 4.562 per cent, 11.780 per cent and 11.322 per cent for area, production and yield, respectively.

Table 4.7: Instability of Area, Production and Yield of Maize Crop in Jammu and Kashmir State

S.No.	*Particulars*	*Period I (1987-88 to 1996-97)*	*Period II (1997-98 to 2006-07)*	*Overall Period (1987-88 to 2006-07)*
1.	Area	1.850	2.071	4.562
2.	Production	15.180	7.322	11.780
3.	Yield	14.612	11.090	11.322

Instability in area, production and yield of maize crop in Jammu region level has been presented in Table 4.8. The Table revealed that during the period I, the variability in area, production and yield was 2.300 per cent, 11.794 per cent and 10.641 per cent, respectively. From the data, it was further observed that the variability in area, production and yield was 1.584 per cent, 8.732 per cent and 9.330 per cent respectively for period II, whereas the overall variability in area was 5.041 per cent, which was minimum as compared to yield (10.20 per cent) and production (10.14 per cent).

Table 4.8: Instability in Area, Production and Yield of Maize Crop in Jammu Region of J&K State

S.No.	*Particulars*	*Period I (1987-88 to 1996-97)*	*Period II (1997-98 to 2006-07)*	*Overall Period (1987-88 to 2006-07)*
1.	Area	2.300	1.584	5.041
2.	Production	11.794	8.732	10.142
3.	Yield	10.641	9.330	10.201

Instability using coefficient of variation of area, production and yield under maize crop for selected districts were shown in Table 4.9. The results indicated that during the period I, the instability in area for Doda and Udhampur districts was 3.941 per cent, and 3.596 per cent, respectively with production variation of 8.260 per cent and 9.072 per cent, respectively and yield variation of 8.891 per cent and 8.645 per cent, respectively. For the period II, variability in area, production and yield was that of 2.363 per cent, 30.861 per cent and 30.121 per cent, respectively for Doda district and for Udhampur district, it was 4.520 per cent, 17.836 per cent and 18.207 per cent, respectively. The overall instability for the area, production and yield for Doda and Udhampur was 4.370 per cent, 28.150 per cent, 29.601 per cent and 9.772 per cent, 13.812 per cent, 15.060 per cent, respectively.

Table 4.9: Instability of Area, Production and Yield of Maize Crop in Selected Districts

S.No.	*Particulars*	*Period I (1987-88 to 1996-97)*	*Period II (1997-98 to 2006-07)*	*Overall Period (1987-88 to 2006-07)*
	Doda			
1.	Area	3.941	2.363	4.370
2.	Production	8.260	30.861	28.150
3.	Yield	8.891	31.121	29.601
	Udhampur			
1.	Area	3.596	4.520	9.772
2.	Production	9.072	17.836	13.812
3.	Yield	8.645	18.207	15.060

The instability in area, production and yield of maize crop for selected maize growers of Doda district for the year 2007-08 is given in Table 4.10. The results pertaining to the variability in acreage was highest for medium farmers (56.923 per cent) and lowest for large farmers (30.960 per cent). While as instability in area for small and marginal farmers was 37.175 per cent and 46.084 per cent, respectively. The table further revealed that the variability for maize production was found to be highest for marginal farmers (48.243 per cent) followed by small farmers (43.409 per cent), then medium farmers (39.053 per cent) and large farmers (34.782 per cent). It was also observed that the medium farmers had highest instability for yield, which was 47.402 per cent. Whereas, instability for marginal, small and large farmers was 43.451 per cent, 31.225 per cent and 29.086 per cent, respectively.

Table 4.10: Instability of Area, Production and Yield of Selected Maize Growers in Doda District for the year 2007-08

S.No.	*Particulars*	*Marginal*	*Small*	*Medium*	*Large*	*Overall*
1.	Area	46.084	37.175	56.923	30.960	41.650
2.	Production	48.243	43.409	39.053	34.782	44.214
3.	Yield	43.451	31.225	47.402	29.086	37.736

Estimates of instability in area, production and yield of maize for selected maize growers of Udhampur district has been presented in Table 4.11. The Table indicated that the variability in area, production and yield for marginal farmers was 53.824 per cent, 54.053 per cent and 76.654 per cent, respectively. From the data, it was further observed that the variability in area, production and yield was 51.583 per cent, 63.207 per cent, 67.490 per cent, respectively for small farmers, 49.583 per cent, 63.247 per cent, 67.419 per cent, respectively for medium farmers, whereas the variability in area, production and yield of maize crop for large farmers was 45.031 per cent, 47.309 per cent and 54.479 per cent, respectively.

Table 4.11: Instability of Area, Production and Yield of Selected Maize Growers in Udhampur District for the year 2007-08

S.No.	*Particulars*	*Marginal*	*Small*	*Medium*	*Large*	*Overall*
1.	Area	53.824	51.583	49.583	45.031	51.012
2.	Production	54.053	63.207	63.247	47.309	57.184
3.	Yield	76.654	67.490	67.419	54.479	69.898

4.9.4 Decomposition of Maize Crop Production into Area, Yield and Interaction Effects

An analysis of growth rate and instability in area, production and yield of maize indicated the general pattern of growth and variability in maize crop. But, this does not evaluate the exact contribution of area and yield to production. Therefore, an attempt was made to determine the extent of relative contribution of area, yield and their interaction on the change in production of maize crop between the two periods for India (1987-88 to 1996-97 and 1997-98 to 2006-07), which was done by decomposing the crop production as shown in Table 4.12. The data indicated that during period I, the change in area brought about a change in production by 0.714 million tonnes while keeping the productivity constant. Similarly, a change in yield changed the production by 3.892 million tonnes keeping area constant and the interaction effect of area and yield was 2.778 million tonnes on production during the period I. The data further revealed that during period II, the change in area and yield brought about a change in production by 2.470 million tonnes and 0.442 million tonnes, respectively. The interaction effect of area and yield was 1.091 million tonnes on production during the period II. But the overall effect of two periods on production was 2.225 million tonnes and 4.225 million tonnes with respect to area and yield, respectively.

Table 4.12: Effect of Change in Area, Yield and Interaction of Differential Production of Maize Crop in India

(Million tonnes)

Particulars	*Area Effect* $Y_0 \Delta A$	*Yield Effect* $A_0 \Delta Y$	*Interaction Effect* $\Delta A \Delta Y$
1987-88 to 1996-97	0.714	3.892	2.778
1997-98 to 2006-07	2.470	0.442	1.091
Overall Total 1987-88 to 2006-07	2.225	4.225	9.40

The area, yield and their interactional effect on the change in production of maize crop between the two periods (1987-88 to 1996-97 and 1997-98 to 2006-07) for Jammu and Kashmir State is shown in Table 4.13. The Table indicated that during period I, the change in area brought about a change in production by 0.012 million tonnes while keeping yield constant. Similarly, while keeping area constant yield effect was 0.014 million tonnes. Area and yield interaction effect was 0.005 million tonnes on production during the period I. Following the similar pattern, during the period II, the change in area brought about a change in production by 0.014 million tonnes and yield by 0.027 million tonnes. The data further indicated that the interaction effect of area and yield was 0.009 million tonnes on production during the period II and the overall change in production brought by change in area and yield was 0.303 million tonnes and 0.142 million tonnes, respectively. The interaction effect of area and yield for overall period (1987-88 to 2006-07) was 0.014 million tonnes.

Table 4.13: Effect of Change in Area, Yield and Interaction of Differential Production of Maize Crop in Jammu and Kashmir State

(Million tonnes)

Particulars	*Area Effect* $Y_0 \Delta A$	*Yield Effect* $A_0 \Delta Y$	*Interaction Effect* $\Delta A \Delta Y$
1987-88 to 1996-97	0.012	0.014	0.005
1997-98 to 2006-07	0.014	0.027	0.009
Overall Total 1987-88 to 2006-07	0.303	0.142	0.014

The decomposition of maize production in Jammu region is presented in Table 4.14. The results clearly showed that the area contributed positively (0.018 million tonnes) in production during the period I, whereas the change in yield had negatively affected production and had reduced production by (-) 0.030 million tonnes. The interaction effect of area and yield had reduced production by 0.001 million tonnes during period I. Where as during period II, change in area, yield and interaction effect of both area and yield had a positive effect on production *i.e.* 0.017 million tonnes, 0.004 million tonnes and 0.001 million tonnes, respectively. For the overall period the area effect was 0.056 million tonnes on production and that of yield was (-) 0.026 million tonnes. Their (area and yield) interactive effected had decreased production by 0.004 million tonnes.

Table 4.14: Effect of Change in Area, Yield and Interaction of Differential Production of Maize Crop in Jammu Region of J&K State

(Million tonnes)

Particulars	*Area Effect* $Y_0 \Delta A$	*Yield Effect* $A_0 \Delta Y$	*Interaction Effect* $\Delta A \Delta Y$
1987-88 to 1996-97	0.018	(-) 0.030	(-) 0.001
1997-98 to 2006-07	0.017	0.004	0.001
Overall Total 1987-88 to 2006-07	0.056	(-) 0.026	(-)0.004

Table 4.15 presented decomposition of area and yield change on production of maize crop for sample districts as well. The data revealed that in Doda district, the area effect contributed positively in increasing the production in period I (0.008 million tonnes), period II (0.004 million tonnes) and overall period (0.010 million tonnes), but the yield effect and

interaction effect of area and yield was not an encouraging one and reduced production by 0.013 million tonnes and 0.001 million tonnes during period I and 0.024 million tonnes and 0.001 million tonnes in period II, respectively. During overall period, the production decreased by 0.010 million tonnes and 0.021 million tonnes, respectively due to yield and interaction effect. The Table further revealed that during the period I, change in area brought about a change in production by 0.001 million tonnes and change in productivity by 0.009 million tonnes for Udhampur district. The interaction effect of area and productivity was 0.002 million tonnes on production. Even period II, the area effect and interaction effect positively contributed to production by 0.012 million tonnes and 0.002 million tonnes, respectively, whereas, the yield effect was negative and reduced production by 0.016 million tonnes during period II. For the overall period, (1987-88 to 2006-07) productivity effect (0.024 million tonnes) and interaction effect (0.007 million tonnes) showed negative effect on production whileas, area effect had shown a positive effect on production by 0.027 million tonnes.

Table 4.15: Effect of Change in Area, Yield and Interaction of Differential Production of Maize in Selected Districts

(Million tonnes)

Particulars	*Area Effect* $Y_0 \Delta A$	*Yield Effect* $A_0 \Delta Y$	*Interaction Effect* $\Delta A \Delta Y$
		Doda	
1987-88 to 1996-97	0.008	(-) 0.013	(-) 0.001
1997-98 to 2006-07	0.004	(-) 0.024	(-) 0.001
Overall Total 1987-88 to 2006-07	0.010	(-) 0.010	(-) 0.021
		Udhampur	
1987-88 to 1996-97	0.001	0.009	0.002
1997-98 to 2006-07	0.012	(-) 0.016	0.002
Overall Total 1987-88 to 2006-07	0.027	(-) 0.024	(-) 0.007

4.9.5 Adoption of New Maize Technology

The adoption of new maize technology means using the entire package of practices for cultivation along with improved variety. To capture this, the adoption index was calculated for individual farmers which included adoption of improved variety, adoption of recommended doses of chemical fertilizers, recommended number of irrigation etc. as discussed in methodology chapter and is given in Table 4.16. It was observed that 49 per cent of the farmers fell under the category of low adoption (less than 33 per cent adoption index), 40 per cent belonged to the category of medium adoption (33-66 per cent adoption index) and only 11 per cent belonged to high adoption (more than 66 per cent adoption index).

Table 4.16: Distribution of Maize Growers under different Levels of Adoption of New Maize Technology

Adoption Level	*Extent of Adoption (Per cent of farmers)*
Low adopters (< 33 per cent TAI)	49.00
Medium adopters (33- 66 per cent TAI)	40.00
High adopters (> 66 per cent TAI)	11.00

TAI: Technological Adoption Index.

4.9.6 Cultivar-wise Area under Maize

The distribution of area under different maize cultivars on the sample farms is represented in Table 4.17. Out of the total cropped area, marginal, small, medium and large farmers were allocating 0.36 hectare, 0.89 hectare, 1.13 hectare and 1.59 hectare, respectively to maize crop on an average. The data further indicated that about 15.48 per cent, 19.67 per cent, 43.44 per cent and 32.65 per cent area under maize was occupied by hybrid cultivars in Marginal, small, medium and large farmers, respectively and overall about 21.78 per cent was occupied under this variety. The traditional varieties occupied maximum area under its cultivation as compared to hybrid varieties, it was highest for marginal farmers (84.52 per cent) followed by small (80.33 per cent) than by large farmers (67.35 per cent) and lowest for medium farmers (56.56 per cent). On an average the area under traditional varieties in per cent of total area was 78.22 per cent. This revealed that majority of farmers were cultivating traditional cultivars of maize crop in study area.

Table 4.17: Distribution of Area under Maize Cultivar on Sample Farms

(in per cent of total maize area)

S.No.	*Cultivar*	*Marginal*	*Small*	*Medium*	*Large*	*Overall*
1.	Hybrid	15.48	19.67	43.44	32.65	21.78
2.	Traditional	84.52	80.33	56.56	67.35	78.22
	Total	100.00 (0.36)	100.00 (0.89)	100.00 (1.13)	100.00 (1.59)	100.00 (0.72)

Figures in parenthesis indicate total area under maize per hectare.

4.9.7 Sources of Seed Material of Maize Crop

The source of maize seed used by sample farmers is given in Table 4.18. The table indicated that maximum farmers were using their own farm produced seed *i.e.,* it was 61.08 per cent for medium farmers with highest percentage followed by small farmers (56.54 per cent), 55.55 per cent for marginal and 50.78 per cent for large farmers with 56.02 per cent on an average. Farmers who brought the maize seed of maize from other farmers were 18.64 per cent, 19.39 per cent, 13.79 per cent and 28.53 per cent for marginal, small, medium and large farmers, respectively. It was also observed from the Table 4.18 that about 24.13 per cent of the medium farmers brought their seed from the dealers followed by large farmers (20.66 per cent) and then by marginal farmers (20.53 per cent) and lowest by the small farmers *i.e.,* 16.13 per cent. The farmers in sample area were not getting the seed from government agencies, it was only 5.28 per cent, 7.94 per cent, 1.00 per cent and 0.03 per cent for the marginal, small,

Table 4.18: Sources of Seed Materials of Maize Used by the Sample Farmers, 2007-08

S.No.	*Source*	*Marginal*	*Small*	*Medium*	*Large*	*Overall*
1.	Own seed	55.55	56.54	61.08	50.78	56.02
2.	Other farmers	18.64	19.39	13.79	28.53	19.25
3.	Dealer	20.53	16.13	24.13	20.66	19.71
4.	Govt. agency	5.28	7.94	1.00	0.03	5.02
	Total	100.00 (14.88)	100.00 (22.13)	100.00 (32.04)	100.00 (36.40)	100.00 (21.18)

Figures in parenthesis indicate total maize seed (Kg/ha).

medium and large farmers, respectively. The average maize received from government agencies was merely 5.02 per cent.

The data further indicated that large farmers used highest seed rate, it was 36.40 Kg/ha followed by medium (32.04 Kg/ha), small (22.13 Kg/ha) and 14.88 Kg/ha by marginal farmers and average seed rate was 21.18 Kg/ha for the sample farmers in the study area.

4.9.8 Economics and Impact of Improved Maize Technology over Traditional Maize Variety

The results regarding the economics and impact of improved maize technology is given in Table 4.19. The data revealed that per hectare cost of cultivation was the highest for hybrid cultivars (₹6748.63/ha) as compared to traditional cultivars (₹4685.60/ha), but the net returns was more for hybrid cultivars than traditional one *i.e.*, ₹8737.37/ha and ₹5472.40/ha, respectively. The per hectare output value was ₹15486.00 for hybrid and ₹10158.00 for traditional cultivars. The data further indicated that the output-input ratio of hybrid maize (2.29) was also greater than traditional (2.16) cultivars. The analysis of impact of improved maize cultivars on yield as well as on the cost of production is also indicated by Table 4.19. The production technology can be said improved if it either increases the per unit quantity of production with the given inputs or enables to reduce the cost of production (per unit of output). Table indicated that maize yield had increased from 16.93 qtls/ha (traditional) to 25.81qtls/ha with hybrid cultivar. In other words, the maize yield had increased by 52.45 per cent with the use of hybrid cultivars over traditional

Table 4.19: Economics and Impact of Improved Maize Technology over Traditional Maize Variety

Sl.No.	*Particulars*	*Traditional*	*Hybrid*
1.	Input cost (Rs/ha)	4685.60	6748.63
2.	Output value (Rs/ha)	10158.00	15486.00
3.	Net return (Rs/ha)	5472.40	8737.37
4.	Output/input ratio	2.16	2.29
5.	Yield (qtls/ha)	16.93	25.81
6.	Increase in yield (per cent)	–	52.45
7.	Cost of production (Rs/qtl)	338.52	252.32
8.	Reduction in cost (per cent)	–	25.46

variety. The per quintal cost of production had reduced to ₹252.32 for hybrid than ₹338.52 for the traditional variety. In other words, the cost reduced by 25.46 per cent for hybrid varieties when compared to traditional varieties.

4.9.9 Socio-economic Constraints to Maize Production by Farmers

There are so many problems faced by the farmers in adoption of improved maize technology is given in Table 4.20. 83 per cent of the farmers in the sample area opined that the adoption of maize technology is expensive. 96.21 per cent and 98.65 per cent of the farmers were not using the recommended quantity of micro-nutrients and pesticides because of their high cost. In addition to this there were 25.10 per cent of farmers who were facing the problem of timely availability of labour while 33.36 per cent opined that labour cost was high. The Table further indicated that improved maize technology was not adopted by the maximum farmers due to the many more problems like small holdings, high cost of transportation, markets were not approachable for the sale of produce, educated members go outside and did not take interest in farming and absence of technical knowledge, which accounted for 58.33 per cent, 68.75 per cent, 41.66 per cent, 50.23 per cent, 58.33 per cent and 84.23 per cent, respectively.

Table 4.20: Socio -Economic Constraints and Expectations Perceived by Sample Farmers

Particulars	*Per cent of Respondent*
Adoption of improved technology is expensive	83.00
Labour not available when needed	25.10
Govt. is not taking step for development of agriculture	58.33
High labour cost	33.36
Small holding	68.75
High cost of micronutrients	96.21
High cost of transportation	41.66
Non-availability of market	50.23
High cost of pesticides	98.65
Educated members go outside	58.33
Absence of technical knowledge	84.23

4.10 Resource-use Efficiency of Sample Farmers (Estimates of OLS)

Efficiency is an important factor in productivity growth. In an economy where resources are scarce and opportunities for new technologies are lacking, inefficiency studies will be able to show that it is possible to raise productivity by improving efficiency without increasing the resource base or developing new technology. Estimates of the extent of inefficiency also help in deciding whether to improve efficiency or to develop new technologies to raise agricultural productivity. It is recognized fact that functional analysis of the relationship between output and input factors serves as a powerful and reliable tool for resource allocation in the cultivation of the crop at the farm level. The regression co-efficient of factor input from Cobb-Douglas production function (OLS) were used to calculate the Marginal Value Productivity (MVP) at Geometric Mean level for the average farms. The estimates of Cobb-Douglas production function (OLS) are given in Table 4.21. Yield of maize was regressed on various factors of production *viz.* labour, capital, irrigation and fertilizers (N, P and K). These variables were taken as the explanatory variables. The perusal of the data depicted in Table 4.21 revealed that crop production function for maize with R^2 value at 0.44 was statistically significant meaning that 44 per cent of the total variation in maize production was explained by the independent or explanatory variables under consideration.

Table 4.21: Estimated Values of the Coefficients and Related Statistics of Cobb-Douglas Production Function Model

Explanatory Variables	*Coefficients*	*Standard Error*
Constant	(-)1.050	0.263
Labour (L)	0.338*	0.111
Capital (K)	0.306*	0.115
Irrigation (I)	0.043	0.041
Fertilizers (F)		
N	0.324*	0.063
P	(-)0.097**	0.046
K	0.202*	0.041
R^2	0.44	

* Significant at 1 per cent level of probability.

** Significant at 5 per cent level of probability.

Table 4.21 further indicated that the regression coefficient of labour, capital, fertilizers (N and K) were found to be positively significant at 1 per cent level of probability with their values at 0.338, 0.306, 0.324 and 0.202, respectively. Fertilizer like phosphorus (P) was negatively significant at 5 per cent level of probability with its value as (-) 0.097. The regression coefficient of the irrigation was, positive (0.043) but non-significant.

4.10.1 Allocative Efficiency of Sample Farms

Further, for estimating the marginal conditions for profit maximization, allocative efficiency was computed. To attain the goal of profit maximization *i.e.*, for efficient resources allocation, more use of the resources as long as the value of the added product is greater than cost of the added amount of the resources in producing it. The resources are to be considered efficiently used and profit will be maximum, when the marginal value product (MVP) and marginal factor cost (MFC) for each input is equal. The MVP of a particular resource represents the addition to gross return in value terms caused by an addition of one unit of that resource while other inputs are held constant. The estimated MVP, MFC and allocative efficiency (ratio of MVP to MFC) of different inputs in study area are presented in Table 4.22. The Table indicated that the marginal value productivity of labour, capital, irrigation and fertilizers nitrogen and potash (N and K) were positive with their values at 1.077, 5.242, 9.045, 16.221 and 48.51, respectively while as the marginal value productivity estimated for fertilizer (P) was negative (-12.055) and the marginal factor cost for all variables like labour, capital, irrigation and

Table 4.22: Allocative Efficiency of Sample Farms

Explanatory Variables	*Coefficients*	*MVP*	*MFC*	*Allocative Efficiency*
Constant	(-)1.050			
Labour (L)	0.338	1.077	76.766	0.014
Capital (K)	0.306	5.242	7.846	0.668
Irrigation (I)	0.043	9.045	8.871	1.019
Fertilizers (F)				
N	0.324	16.221	5.000	3.244
P	(-) 0.097	(-) 12.055	6.958	(-)1.732
K	0.202	48.510	3.625	13.380

fertilizers nitrogen, phosphate and potash (N, P and K), 76.766, 7.846, 8.871, 5.000, 6.958 and 3.625, respectively. The allocative efficiency which is the ratio of MVP and MFC was 0.014, 0.668, 1.019, 3.244 and 13.38 respectively for labour, capital, irrigation and fertilizers (N and K), respectively and the allocative efficiency of fertilizer (P) with negative value was (-) 1.732.

4.10.2 Technical Efficiency of Sample Farms

Table 4.23: Results of Maximum Likelihood Estimation for Frontier Production Function

Variables	*Frontier Production Function Estimates (MLE)*	*Standard Error*	*Variance Parameters*
Constant	(-) 0.709	0.243	$\sigma^2_{u=}$ 0.196
Labour (L)	0.378 *	0.081	$\sigma^2_{v=}$ 0.011
Capital (K)	0.336 *	0.116	σ = 0.455
Irrigation (I)	0.225	0.380	λ = 4.219
Fertilizers (F)			
N	0.244 *	0.059	γ = 0.946
P	(-) 0.383	0.042	Likelihood=-24.183
K	0.292*	0.033	

* Significant at 1 per cent level of probability.

Estimation of the efficiency level helps to decide whether to improve the existing efficiency level or to develop new technologies to raise the productivity level. A farm is technically inefficient in the sense that if it fails to produce maximum output from a given level of input *i.e.*, it results into equi-proportionate, over or under utilization of all inputs. Using maximum likelihood estimation techniques, the Stochastic Production Frontier was employed to estimate technical efficiency at farm level in the study area. The dependent variable included in the model was the output of maize crop and the independent variables were labour, capital, fertilizers (N, P and K) and irrigation. The estimates of frontier production function for the maize crop were given in Table 4.23. The table predicted that the regression coefficient for labour (0.378), capital (0.336), fertilizers *i.e.*, N (0.244) and K (0.292) of maize were found to be positive and significant at 1 per cent level of probability and fertilizer (P) was found to be negative

(-0.383) and non significant. The regression coefficient for irrigation (0.225) was positive but non significant. From the Table, it was further observed that in case of variance parameters, the value of lambda (λ) was 4.219 and sigma (σ) was 0.455, which were significantly different from zero indicating a good fit and the correctness of the distributional assumptions specified. The estimated value of sigma-squared (σ^2_u) and sigma-squared (σ^2_v) was 0.196 and 0.011 and the value of gamma (γ) was 0. 946. Log likelihood functions (-24.183) of maize was large and significantly different from zero, indicated a good fit and the correctness of the specific distribution assumption.

Table 4. 24: Elasticity of Production of Different Inputs and their Return to Scale

Variable	*Elasticity*
Labour (L)	0.378
Capital (K)	0.336
Irrigation (I)	0.225
Fertilizers (F)	
N	0.244
P	(-) 0.383
K	0.292
RTS	1.092

Note: RTS denotes returns to scale.

The estimated elasticities of the explanatory variables and returns to scale (RTS) is given in Table 4.24. The data revealed that elasticity values of all the input variables like labour (0.378), capital (0.336), irrigation (0.225), nitrogen (0.244) and potash (0.292) were positive while as the value of phosphate (-0.383) was negative. The analysis of results showed that the return to scale (RTS) was 1.092.

4.10.3 Efficiency Level of Sample Farms

The decile range of frequency distribution of technical efficiency in which the farmer falls is called efficiency level of that farmer. The distribution of farm specific technical efficiency coefficients under the efficiency categories is given in Table 4.25 and Figure 4.4. The minimum technical efficiency was 8 per cent and mean technical efficiency was 52

22.12 qtls/ha (80-90 range) and 25.64 qtls/ha (above 90 range) for farmers having their respective level of technical efficiency.

Table 4.26: Mean Actual and Potential Yield of Maize Crop in Different Efficiency Levels

Categorized Technical Efficiency	*Per cent Number of Farms*	*Technical Efficiency*	*Yield (qtls/ha)*	*Potential Yield (qtls/ha)*
Less than 40	23.75	24.57	7.53	13.98
40-50	20.41	42.11	9.31	15.09
50-60	14.16	50.93	9.46	15.70
60-70	12.08	61.04	10.66	17.47
70-80	10.41	78.80	11.02	18.57
80-90	10. 03	85.47	12.90	22.12
Above 90	9.16	94.77	14.88	25.64
Overall Mean	100	52.66	10.82	18.36

4.10.5 Input Use and Technical Efficiency

Input use varies across range of technical efficiency which is shown in Table 4.27. It was observed that most efficient producers (technical efficiency greater than 90) used more inputs that producers who were technically least efficient. The technically most efficient producers (above 90 per cent) had the highest average yield 14.88 qtls/ha and least efficient farmers (less than 40 per cent) 7.53 qtls/ha. The combination of inputs

Table 4. 27: Input Uses and Technical Efficiency

Categorized Technical Efficiency	*Yield (qtls/ha)*	*Fertilizers (Kg/ha)*			*Seed (Kg/ha)*	*Man Days/ha*
		(N)	*(P)*	*(K)*		
Less than 40	7.53	23.08	7.01	4.79	20.58	50.12
40-50	9.31	31.33	8.13	5.19	21.32	48.06
50-60	9.46	36.43	8.22	7.07	23.16	41.73
60-70	10.66	37.73	8.65	9.03	24.94	43.33
70-80	11.02	38.86	10.37	9.77	26.18	45.30
80-90	12.90	42.38	10.76	10.15	30.05	42.53
Above 90	14.88	46.44	11.56	12.20	32.29	40.78
Overall Mean	10.82	36.60	9.24	8.31	25.50	44.55

used by most efficient producers was N, P, K (46.44 Kg/ha, 11.56 Kg/ha and 12.20 Kg/ha), seed (32.29 Kg/ha), labour 40.78 mandays/ha for technically efficient and on the other hand, 23.08 Kg/ha, 7.01 Kg/ha, 4.79 Kg/ha, 20.58 Kg/ha and 50.12 mandays/ha, respectively were used by the least efficient farmers. The data further revealed that the farmers in efficiency level of 40-50 per cent used the combination of inputs 31.33 N Kg/ha, 8.13 P Kg/ha, 5.19 K Kg/ha, 21.32 Kg/ha seed, and 48.06 mandays/ha human labour. Average combination of various inputs used was 36.60 Kg/ha for N, 9.24 Kg/ha for P, 8.31 Kg/ha for K, 25.50 Kg/ha for seed and 44.6 mandays/ha for human labour.

4.11 Factors Affecting Technical Efficiency

The measure of technical efficiency of a farm indicated that if any farm is successful in converting all the physical units into output and efficiency of converting is equal to the hypothetical frontier production function, then it is said to be an efficient farm and if any farm falls short of this requirement then the farm is termed as technically inefficient farm. The technical efficiency of many farmers is determined by various factors like age of the head, education, size of farm and proportion of female workers in the family and the same has been presented in Table 4.28. The regression analysis of various factors predicted the regression coefficient for education as 0.023 and farm size as 0.878 which were positive and significant at 1 per cent level of probability. Similarly, the regression coefficient for the proportion of female family workers in the family was positively (0.062) significant at 5 per cent level of probability. The regression

Table 4.28: Determinants of Technical Efficiency in Maize Production

Variable	*Coefficient*	*Standard Error*
Constant	0.164	0.141
Age	(-)0.001	0.002
Education	0.023*	0.011
Female workers	0.062**	0.034
Children	0.036	0.040
Farm size	0.878*	0.004

* Significant at 1 per cent level of probability.

** Significant at 5 per cent level of probability.

coefficient for age of the head was negative (-0.001), but for the children (0.036) was positive and insignificant. Farm size had the highest impact on technical efficiency of a farmer with 0.878 regression coefficient and was significant at 1 per cent level of significance.

4.12 Economics (Costs and Returns) on Maize Production

The economics of maize production included the average operational as well as fixed cost on the basis of various cost concepts and the returns from maize cultivation which are presented under following sub-heads:

4.12.1 Cost Structure for Maize Cultivation in Doda District

Relative profitability of different crops is essential for decision making of farmers about a particular crop. For financial analysis of different enterprises, it is necessary to workout costs of inputs, which needs to be deducted from the value of output. Farmers in the study area used purchased as well as home produced inputs. The costs of home produced inputs were calculated in monetary terms on the basis of opportunity cost principle (*i.e.*, on market price). Table 4.29 presented the operational and fixed costs for maize crop cultivation in the sample farms in Doda district. The perusal of the data indicated that per hectare total cost (operational cost + fixed cost) of maize crop cultivation was ₹13287.60/ha for marginal farmers, ₹12863.35/ha for small farmers, ₹13187.24/ha for medium and ₹12792.22/ha for large farmers, with the overall average of ₹12882.83/ha. From the data it was further observed that the operational cost for marginal farmers (₹6925.78/ha) was highest followed by medium farmers (₹6858.85/ha), large farmers (₹6687.76/ha) and lowest for small farmers (₹6682.56/ha), which constituted 52.12 per cent, 52.01 per cent, 52.27 per cent and 51.95 per cent, respectively out of the total cost. The overall average operational cost was ₹6708.14/ha (*i.e.*, 52.07 per cent of the total cost). Among the input costs the labour cost was the major cost in the operational cost and was relatively higher for marginal farmers (₹4311.96/ha) followed by medium farmers (₹4280.98/ha), large farmers (₹4217.18/ha) and lowest for small farmers (₹4196.00/ha). Out of the total operational cost incurred, the overall average labour cost worked out to be ₹4193.13/ha *i.e.*, 62.50 per cent out of the total operational cost and 32.54 per cent out of the total cost. The Table also revealed the fact that fertilizers and manures occupied the second place after labour cost in the operational

Table 4.29: Cost Structure of Maize Cultivation in Doda District

(Rs/ha)

Items	*Marginal*	*Small*	*Medium*	*Large*	*Overall Average**
A. Operational Cost					
Hired labour	1387.72	1317.54	1370.68	1314.96	1334.84
Family labour	2924.24	2878.46	2910.30	2902.22	2858.29
Total	4311.96	4196.00	4280.98	4217.18	4193.13
Bullock labour	271.45	230.90	261.28	263.48	252.84
Seed	762.67	760.99	751.62	746.97	747.43
Fertilizers and manures	1105.60	1039.66	1099.51	1007.77	1057.02
Interest on working capital	256.00	238.01	245.23	242.12	244.22
Miscellaneous expenditure**	218.10	217.00	220.23	210.24	213.50
Total	6925.78	6682.56	6858.85	6687.76	6708.14
B. Fixed Cost					
Rental value of owned land	4695.23	4626.57	4695.75	4554.31	4582.34
Depreciation on implements and farm buildings	184.12	176.23	183.85	181.60	178.30
Interest on fixed capital (excluding land)	1482.47	1377.99	1448.79	1368.55	1414.05
Total	6361.82	6180.79	6328.39	6104.46	6174.69
Total Cost (A+B)	**13287.60**	**12863.35**	**13187.24**	**12792.22**	**12882.83**

* Weighted average.

** It includes expenses on purchase of small farm implement, transportation cost etc. from agriculture dept./center or market to home.

cost. It was ₹1105.60 for marginal farmers, ₹1039.66 for small farmers, ₹1099.51 for medium farmers and ₹1007.77 for large farmers with an overall weighted average of ₹1057.02. As far as size of holding was concerned, marginal farmers had to bear highest fertilizer and manure cost. The chronology of the cost revealed that seed cost amounted to ₹762.67/ha, had a third place and it was ₹760.99/ha, ₹751.62/ha and ₹746.97/ha for marginal, small, medium and large farm size. For bullock labour, it was ₹271.45/ha for marginal farmers, whereas for other categories of farm size, it was ₹230.90/ha for small farm size and ₹261.28/ha for medium and ₹263.48 for large farm size. The Table further indicated that the interest on working capital was 3.69 per cent for marginal farmers, 3.56 per cent for small, 3.57 per cent for medium and was 3.62 per cent for large farm out of their respective operational cost. As far as the fixed cost was concerned, rental value of owned land, interest on fixed capital and depreciation on fixed capital were the major components of this cost. The data indicated that in fixed cost, rental value of owned land involved highest per hectare cost for all the holdings and was ₹4695.23, ₹4626.57, ₹4695.75 and ₹4554.31 for marginal, small, medium and large farmers, respectively. The per hectare interest on fixed capital was highest for marginal farmers (₹1482.47) followed by medium (₹1448.79), small (₹1377.99), and large farmers (₹1368.55) and the depreciation on implements and buildings was ₹184.12 for marginal farm size, ₹176.23 for small farmers, ₹183.85 for medium farmers and ₹181.60 for large farmers. On an average, the per hectare rental value of owned land, depreciation and interest on fixed capital was ₹4582.34, ₹178.30 and ₹1414.05, respectively and accounted for 74.21 per cent, 2.88 per cent and 22.90 per cent, respectively out of total fixed average cost.

4.12.2 Returns from Maize Production in Doda District

The returns from maize production under the study area of Doda district is presented in Table 4.30. For working out the economics it became necessary to work out costs involved, gross income, net income and benefit cost ratio based on various cost concepts. The Table revealed that per hectare yield was highest for marginal farmers followed by small, medium and large farmers *i.e.*, 20.01 qtls/ha, 19.68 qtls/ha, 19.28 qtls/ha and 19.02 qtls/ha, respectively and the overall average yield was 19.44 qtls/ha. It was further observed that gross income from maize crop was highest for marginal farmers (₹18009.00/ha) followed by small (₹17712.00/ha), medium (₹17352.00/ha) and large farmers (₹17118/ha), respectively. The overall average gross income per hectare was worked out to be ₹17500.27/ha. From the Table, it is evident that both A_1 and A_2 incurred same cost of ₹8826.76/ha, ₹8380.56/ha, 8668.59/ha and ₹8236.56/ha for marginal, small, medium and large farm sizes, respectively because the farmers were not leasing in the land for maize cultivation. The data further indicated that marginal farmers had incurred highest A_1, A_2 costs with lowest cost

Table 4.30: Economics of Maize Production in Doda District

Particulars	*Marginal*	*Small*	*Medium*	*Large*	*Overall Average**
Yield per ha (qtls)	20.01	19.68	19.28	19.02	19.44
Gross income (Rs/ha)	18009	17712	17352	17118	17500.27
			Cost (Rs/ha)		
A_1	8826.76	8380.56	8668.59	8236.56	8476.34
A_2	8826.76	8380.56	8668.59	8236.56	8476.34
B_1	9026.76	8581.16	8968.59	8492.98	8683.36
B_2	9226.76	8912.06	9168.59	8872.98	8937.98
C_1	13037.60	12632.44	13003.71	12553.20	12645.90
C_2	13287.60	12863.35	13187.24	12792.22	12882.83
C_3	14616.36	14149.69	14505.96	14071.44	14171.11
			Net income over above cost concepts (Rs/ha)		
A_1	9182.24	9331.44	8683.41	8881.44	9023.93
A_2	9182.24	9331.44	8683.41	8881.44	9023.93
B_1	8982.24	9130.84	8383.41	8625.02	8816.90
B_2	8782.24	8799.94	8183.41	8245.02	8562.28
C_1	4971.40	5079.56	4348.29	4564.80	4854.36
C_2	4721.40	4848.65	4164.76	4325.78	4617.44
C_3	3392.64	3562.31	2846.04	3046.56	3329.16
			Benefit-cost ratio		
A_1	2.04	2.11	2.00	2.07	2.09
A_2	2.04	2.11	2.00	2.07	2.09
B_1	1.99	2.06	1.93	2.01	2.04
B_2	1.95	1.98	1.89	1.92	1.98
C_1	1.38	1.40	1.33	1.36	1.40
C_2	1.35	1.37	1.31	1.33	1.35
C_3	1.23	1.25	1.19	1.21	1.23

* Weighted average.

A_2: There was not a single farmer who leased in land for maize cultivation.

B_2: The information regarding rental value of owned land was collected from revenue department of J&K State.

C_1: Imputed value of family labour was based on the per day wages of hired labour.

C_3: C_2 + 10 per cent of C_2 for the managerial and the entrepreneurial functions performed by the farmer.

for large farmers. It was also observed from the Table, that marginal farmers had also incurred highest B_1 costs of ₹9026.76/ha, again with lowest cost of B_1 for large farmers (₹8492.98/ha). For small farmers, the B_1 cost involved was ₹8581.16/ha, for medium farmers, it was ₹8968.59/ha. As far as B_2 cost was concerned, the Table indicated that marginal farmer had again to bear highest cost of ₹9226.76/ha, followed by medium farmer (₹9168.59/ha), small farmer (₹8912.06/ha) and that of large farmer (₹8872.98/ha). Similar pattern was followed with respect to C_1 and C_2 cost. It was ₹13037.60/ha, ₹12632.44/ha, ₹13003.71/ha and ₹12553.20/ha for marginal, small, medium and large farmers, respectively for C_1 cost. As for C_2 cost was concerned, it was ₹13287.60/ha (marginal farmers), ₹12863.35/ha (small farmers), ₹13187.24 (medium farmers) and ₹12792.22/ha (large farmers). On an average ₹8476.34/ha (A_1 and A_2), ₹8683.36/ha (B_1), ₹8937.98/ha (B_2), ₹12645.90/ha (C_1) and ₹12882.83/ha (C_2), where as C_3 was ₹14616.36/ha, ₹14149.69/ha, ₹14505.96/ha and ₹14071.44/ha, for marginal, small, medium and large farmers, respectively. The perusal of the data further indicated that the net income over various costs with respect to A_1 and A_2 was highest for small farmers (₹9331.44/ha), followed by marginal farmers (₹9182.24/ha) then by large farmers (₹8881.44/ha) and medium farmers (₹8683.41/ha). Small farmers had highest net income *i.e.*, ₹9130.84/ha and ₹8799.94/ha over cost B_1 and B_2 and lowest for medium farmers (₹8383.41/ha and ₹8183.41/ha). In case of marginal and large farmers, the net income was ₹8982.24/ha, ₹8625.02/ha for B_1 cost and ₹8782.24/ha, ₹8245.02/ha for B_2 cost respectively. For C_1, the net income for marginal, small, medium and large farmers were ₹4971.40/ha, ₹5079.56/ha, ₹4348.29/ha and ₹4564.80/ha, respectively. As far as the net income over C_2 was concerned, it was ₹4721.40/ha, ₹4848.65/ha, ₹4164.76/ha and ₹4325.78/ha and for C_3 cost it was ₹3392.64/ha, ₹3562.31/ha, ₹2846.04/ha and ₹3046.56/ha for marginal, small, medium and large farmers, respectively. The average net income over various costs was positive *i.e.*, it was ₹9023.93/ha, ₹9023.93/ha, ₹8816.90/ha, ₹8562.28/ha, ₹4854.36/ha, ₹4617.44/ha and ₹3329.16/ha, respectively.

The return per rupee invested in maize crop production or benefit-cost ratio were highest in case of cost A_1 and A_2 for small farmers (2.11) followed by large farmers (2.07), marginal farmers (2.04) and lowest for medium farmers *i.e.*, 2.00. The results further showed that small farmers had incurred highest benefit cost ratio on other cost concepts also, for B_1 it was 2.06. While for other size of holdings benefit-cost ratio for B_1 was 1.99, 1.93 and 2.01, respectively for marginal, medium and large farmers. For B_2, it was highest for small farmers (1.98) and lowest for medium farmers (1.89) and for C_1 and C_2 it was also highest for small farmers (1.40 and 1.37) and lowest for medium farmers (1.33 and 1.31) and more than one for all costs. In case of C_3 cost the benefit cost ratio was 1.23, 1.25, 1.19 and 1.21 for marginal, small, medium and large farmers, respectively.

4.12.3 Cost Structure for Maize Cultivation in Udhampur District

The operational and fixed cost for maize crop cultivation in the sample farms of Udhampur district are given in Table 4.31. The data indicated that per hectare total cost of maize crop cultivation was ₹10577.76/ha for marginal farmers, ₹10465.12/ha for small farmers, ₹10202.08/ha for medium, ₹10178.73/ha for large farmers. The average cost of cultivation worked out to be ₹10430.84/ha. From the Table it was observed that like Doda district, the labour cost was the major item in operational cost and worked out to be ₹3908.08/ha for marginal farmers followed by small farmers (₹3815.33/ha), medium farmers (₹3708.18/ha) and lowest for large farmers (₹3673.28/ha) with the overall average of ₹3818.89/ha. Labour cost was followed by cost incurred on fertilizers and manure, which worked out to be ₹1339.29/ha for marginal farmers, ₹1328.02/ha for small farmers, ₹1327.08/ha for marginal farmers, ₹1325.32 for large farmers with an overall average of ₹1332.48/ha, which accounted for 21.26 per cent, 21.32 per cent, 21.66 per cent and 21.93 per cent, respectively of total operational cost. Cost on seed was the third highest among various operational costs. It worked out to be ₹382.97/ha (marginal farmers), ₹389.15/ha (small farmers), ₹385.21/ha (medium farmers), and ₹378.28/ha (large farmers) with an overall average of ₹384.33/ha. Furthermore, interest on working capital was ₹230.01/ha (marginal farmers), ₹269.00/ha (small farmers), ₹288.19/ha (medium farmers) and ₹278.10/ha (large farmers), which accounted to be 3.65 per cent, 4.31 per cent, 4.70 per cent and 4.58 per cent for marginal, small, medium and large farmers, respectively out of the total operational cost for different

Table 4.31: Cost Structure of Maize Cultivation in Udhampur District

(Rs/ha)

Items	*Marginal*	*Small*	*Medium*	*Large*	*Overall Average**
A. Operational Cost					
Hired Labour	1708.48	1716.93	1699.38	1670.00	1704.11
Family Labour	2199.60	2098.40	2008.80	2003.28	2114.78
Total	3908.08	3815.33	3708.18	3673.28	3818.89
Bullock Labour	246.38	235.80	216.32	215.90	234.41
Seed	382.97	389.15	385.21	378.28	384.33
Fertilizers and Manures	1339.29	1328.02	1327.08	1325.32	1332.48
Interest on Working Capital	230.01	269.00	288.19	278.10	259.53
Miscellaneous Expenditure**	190.63	191.00	190.25	189.10	190.21
Total	6297.36	6227.30	6125.27	6070.07	6219.85
B. Fixed Cost					
Rental Value of Owned Land	3097.31	3043.09	2932.60	2921.64	3031.59
Depreciation on Implements and Farm Buildings	186.00	198.91	188.00	189.02	189.98
Interest on Fixed Capital (excluding land)	997.09	995.82	956.21	998.00	989.39
Total	4280.40	4237.82	4076.81	4108.66	4210.96
Total Cost (A+B)	**10577.76**	**10465.12**	**10202.08**	**10178.73**	**10430.84**

* Weighted average.

** It includes expenses on purchase of small farm implement, transportation cost etc. from agriculture dept./center or market to home.

farm sizes. For bullock labour, it was 3.91 per cent (marginal farmers), whereas for other categories of farm size, it was 3.78 per cent (small farmers), 3.53 per cent (medium farmers) and 3.55 per cent (large farmers) out of the total operational cost. The data further indicated that as far as the fixed cost, rental value of owned land involved highest per hectare cost for all the holdings and was ₹3097.31, ₹3043.09, ₹2932.60 and ₹2921.64 for marginal, small, medium and large farmers, respectively.

The per hectare interest on fixed capital was highest for large farmers (₹998.00) followed by marginal (₹997.09), medium (₹995.82) and small farmers (₹956.21) and the depreciation on implements and buildings was lowest for marginal farmers (₹186.00) and highest for small farmers (₹198.91). On an average the per hectare rental value of owned land, depreciation and interest on fixed capital was ₹3031.59, ₹189.98 and ₹989.39, respectively.

4.12.4 Returns from Maize Production in Udhampur District

The returns from maize production under the study area of Udhampur district is presented in Table 4.32. The data revealed that per hectare yield was highest for marginal farmers followed by small, medium and large farmers *i.e.*, 16.98 qtls/ha, 15.21 qtls/ha, 14.87 qtls/ha and 14.68 qtls/ha, respectively and the overall average yield was 15.86 qtls/ha. The gross income from maize crop was highest for marginal farmers (₹15282.00/ha) followed by small (₹13689.00/ha), medium (₹13383.00/ha) and large farmers (₹13212.00/ha), respectively. The overall average gross income per hectare was worked out to be ₹14276.85/ha. Both A_1 and A_2 incurred same cost of ₹7193.50/ha, ₹7001.78/ha, 6707.39/ha and ₹6693.83/ha for marginal, small, medium and large farm sizes, respectively because the farmers were not leasing in the land for maize cultivation and marginal farmers had incurred highest A_1, A_2 costs with lowest cost for large farmers. It was also observed that marginal farmers had incurred highest B_1 cost it was ₹7879.67/ha and lowest for large farmers (₹6820.56/ha). For small farmers the B_1 cost involved was ₹7270.60/ha, for medium farmers, it was ₹6911.47/ha and for large farmers ₹6820.56/ha. As far as B_2 cost was concerned, where as small farmers bear highest cost of ₹8313.60/ha followed by marginal farmers (₹8069.22/ha), medium farmers (₹7826.07/ha) and large farmers (₹7438.36/ha). As far as C_1 cost was concerned marginal farmers bear highest cost, which was ₹9388.22/ha followed by small farmers (₹9022.12/ha), medium farmers (₹9002.02/ha) and large farmers (₹9000.93/ha) and for C_2 cost marginal farmers again with the highest cost of ₹10577.76/ha followed by small farmers with ₹10465.12, medium farmers with ₹10202.08/ha and for large farmers ₹10178.73/ha, whereas, for the $C_{3,}$ the cost involved was ₹11635.54, ₹11511.63, ₹11222.29 and ₹11196.60 for marginal, small, medium and large farmers,

Table 4.32: Economics of Maize Production in Udhampur District

Particulars	Marginal	Small	Medium	Large	Overall Average*
Yield per ha (qtls)	16.98	15.21	14.87	14.68	15.86
Gross income (Rs/ha)3	15282.00	13689.00	13383.00	13212.00	14276.85
			Cost (Rs/ha)		
A_1	7193.50	7001.78	6707.39	6693.83	7008.35
A_2	7193.50	7001.78	6707.39	6693.83	7008.35
B_1	7879.67	7270.60	6911.47	6820.56	7417. 51
B_2	8069.22	8313.60	7826.07	7438.36	8448.54
C_1	9388.22	9022.12	9002.02	9000.93	9209.98
C_2	10577.76	10465.12	10202.08	10178.73	10430.84
C_3	11635.54	11511.63	11222.29	11196.6	11473.92
			Net income over above cost concepts (Rs/ha)		
A_1	8088.50	6687.22	6675.61	6518.17	7268.50
A_2	8088.50	6687.22	6675.61	6518.17	7268.50
B_1	7402.33	6418.40	6471.53	6391.44	6859.35
B_2	7212.78	5375.40	5556.93	5773.64	5828.31
C_1	5893.78	4666.88	4380.98	4211.07	5066.87
C_2	4704.24	3223.88	3180.92	3033.27	3846.01
C_3	3646.46	2177.37	2160.71	2015.39	2802.92
			Benefit-cost ratio		
A_1	2.12	1.95	1.99	1.97	2.03
A_2	2.12	1.95	1.99	1.97	2.03
B_1	1.93	1.88	1.93	1.94	1.92
B_2	1.89	1.64	1.71	1.77	1.68
C_1	1.62	1.51	1.48	1.46	1.55
C_2	1.44	1.30	1.31	1.30	1.36
C_3	1.31	1.18	1.19	1.18	1.24

* Weighted average.

A_2: There was not a single farmer who leased in land for maize cultivation.

B_2: The information regarding rental value of owned land was collected from revenue department of J&K State.

C_1: Imputed value of family labour was based on the per day wages of hired labour.

C_3: C_2 + 10 per cent of C_2 for the managerial and the entrepreneurial functions performed by the farmer.

respectively with an overall average cost of ₹7008.35/ha (A_1 and A_2), ₹7417.51/ha (B_1), ₹8448.54/ha (B_2), ₹9209.98/ha (C_1), ₹10430.84/ha(C_2) and ₹11473.92/ha (C_3). The net income over various costs with respect to A_1 and A_2 was highest for marginal farmers (₹8088.50/ha), followed by small farmers (₹6687.22/ha) then by medium farmers (₹6675.61/ha) and ultimately large farmers (₹6518.17/ha). It was further observed that net income over cost B_1 was highest for marginal farmers (₹7402.33/ha) and lowest for large farmers (₹6391.44/ha). The net return under B_1 cost in case of small farmers was ₹6418.40/ha and medium farms was ₹6471.53/ha, where as for B_2 cost ₹5375.40 for small farmers and ₹5556.93/ha for medium farmers. For C_1, the net income for marginal, small, medium and large farmers were ₹5893.78/ha, ₹4666.88/ha, ₹4380.98/ha and ₹4211.07/ha, respectively. Net income over C_2 was ₹4704.24/ha, ₹3223.88/ha, ₹3180.92/ha and ₹3033.27/ha for marginal, small, medium and large farmers and for C_3 cost it was ₹3646.46, ₹2177.37, ₹2160.71 and ₹2015.39, respectively. The overall average net income over A_1 and A_2, B_1, B_2 C_1, and C_3 was positive *i.e.*, it was Rs 7268.50/ha, ₹6859.35/ha, ₹5828.31/ha, ₹5066.87/ha, ₹3846.01/ha and ₹2802.92/ha, respectively.

The return per rupee invested in maize crop production or benefit-cost ratio were highest in case of cost A_1 and A_2 for marginal farmers (2.12) followed by medium farmers (1.99), large farmers (1.97) and lowest for small farmers *i.e.*, 1.95. The benefit-cost ratio on other cost concepts was 1.93, 1.88, 1.93 and 1.94, respectively for B_1 in case of marginal and small, medium and large farmers. For B_2, it was highest for marginal farmers (1.89) and lowest for small farmers (1. 64) and benefit-cost ratio for C_1 and C_2 was also highest for marginal farmers (1.62 and 1.44 and lowest for large farmers (1.46 and 1.30) and for C_3 it was 1.31, 1.18, 1.19 and 1.18 for marginal, small, medium and large farmers, respectively and more than one for all costs.

4.13 Discussion

4.13.1 Growth Performance of Maize Crop

During the first period (1987-88 to1996-97) there had been impressive positive and significant growth trend in production as well as yield of maize in India (Table 4.1), with their values at 3.637 per cent and 3.531

per cent, respectively. A significant shift occurred during nineties of last century, when winter maize cultivation expanded rapidly in most states of the country due to the new paradigm shift in the seed and price policy of cereals and enhanced support to the export of cereals would bring significant expansion in maize area. Period second (1997-98 to 2006-07) witnessed less growth in yield (0.838 per cent) as compared to period first, while as production trend increased continuously with 3.806 per cent. As far as area was concerned it had increased by 0.859 per cent and 2.658 per cent, respectively during the period I and period II. Area wise, production wise and yield wise, the growth trend of maize crop was 0.870 per cent, 3.132 per cent and 2.299 per cent, respectively for the overall period of 20 years (1987-88 to 2006-07). In this period, the high growth in maize production was mainly attributed from huge expansion in maize area across the country and because of growth in technology coupled with rising demand for the produce. Diversified uses of maize also prompted higher production in the country. Presently, in India, maize is mainly used for preparation of poultry feed and extraction of starch. (Singh *et al.,* 2003 and Mohiuddin *et al.,* 2007).

The compound growth rates were computed for area, production and yield of maize in Jammu and Kashmir State indicated that during period I (1987-88 to 1996-97), maize area expanded significantly at 0.441 per cent, but during period II (1997-98 to 2006-07), it came down to 0.350 per cent, while as for the overall period (1987-88 to 2006-07), it was 0.701 per cent. Similarly, during period I production trend was 3.620 per cent but in period II the growth trend in production decreased to (-) 0.070 per cent significantly. For the overall period, production growth trend was 0.952 per cent. Maize yield recorded a significant growth rate of 3.191 per cent during period I, whereas, it showed a drastic decline of (-) 0.419 per cent in the next period and for the overall period of 20 years, it was 0.252 per cent. Decelerated the maize production and yield has interestingly heavily during the period II, mainly due to decline in area under maize crop during this period, and also on account of higher productivity as well as profitability of the competing crops mainly rice. These findings are similar to the findings of Singh *et al.* (2003).

The compound growth rates for area, production and yield in Jammu region of J&K State (Table 4.3) during the period I (1987-88 to 1996-97), showed positively significant growth rate of 0.633 per cent, 2.511 per cent and 0.155 per cent, respectively. However during period II (1997-98 to 2006-07), the growth rates were positively significant only for area (0.269 per cent) and negatively significant for production (-1.224 per cent) and yield (-1.496 per cent). Decline in production can be attributed to a decline in yield, although the area increased by 0.269 per cent, which showed that yield, is a major contributor to increase maize production. For the overall period (1987-88 to 2006-07), growth rate for area (0.800 per cent) and production (0.351 per cent) was positively significant whereas, for yield it was (-) 0.583 per cent. These results are in conformity with Awasthi, (2003).

As per Table 4.4, district wise growth rates indicated that there was considerable variation in the performance of major maize growing districts of Jammu region during the overall period (1987-88 to 2006-07) in which, Jammu district recorded the highest growth rate of 1.469 per cent in area, followed by Udhampur (1.371 per cent), Doda (0.685 per cent), Rajouri (0.630 per cent), Poonch (0.363 per cent) and Kathua (0.275 per cent). The remarkable growth in production of maize was visible in Jammu district (2.047 per cent), followed by 1.032 per cent for Rajouri and 0.887 per cent for Kathua. The district Udhampur (-0.422 per cent), Doda (-3.637 per cent) and Poonch (-0.727 per cent), experienced negative growth rates for production, which were statistically significant. The growth in yield of maize was positive only for Jammu (0.792 per cent), Kathua (1.558 per cent) and Rajouri district (0.493 per cent), whereas it was negative for Udhampur (-1.462 per cent), Doda (-4.021 per cent) and Poonch district (-0.606 per cent).These findings are supported by Singh *et al.* (2003) and Anupama *et al.* (2005). The results pertaining to the growth performance of maize in major maize growing districts indicated that the rate of growth under maize area was comparatively smaller but positive in all the districts.

Similarly, the growth performance of maize area, production and yield in sample districts of Jammu region in Table 4.5. Observed that the growth rates for production and yield of maize crop in Doda district was not impressive during period I (1987-88 to 1996-97), period II (1997-98 to 2006-07) and overall period (1987-88 to 2006-07), it was (-) 0.360 per cent, (-) 1.007 per cent and (-) 3.637 per cent for production and (-) 0.684 per cent, (-) 1.586 per cent and (-) 4.201 per cent, respectively for yield.

However, area growth rate indicated that there was positive and significant growth in Doda district during these periods. Furthermore, it could be noted that Udhampur district reported positive growth in production (0.179 per cent) and yield (1.435 per cent) during the period I and negative growth trends for area (-0.204 per cent). During the period II the growth rate for area was positive and significant *i.e.* 0.674 per cent and negative for production (-3.441 per cent) and yield (-4.087 per cent), and during the overall period, there was negative and significant growth rates for production (-0.422 per cent) and yield (-1.462 per cent), while positive and significant for area (1.371 per cent). Awasthi (2003) expressed similar views in his study on maize production in Madhya Pradesh, while taking a district level analysis. The limiting factors for maize farming were found to labour problems, erratic rainfall, poor genotypes, incidences of diseases, lack of transportation facilities and low market price.

4.13.2 Instability in Maize Crop Cultivation

The instability of maize crop in India with respect to area, production and yield (Table 4.6) was highest for production, 14.89 per cent in period I and 13.35 per cent in period II among the three variables, for area it was 3.07 per cent and 8.40 per cent, respectively during the period I and II and for yield it was 13.38 per cent and 6.48 per cent, respectively with the overall instability of 9.55 per cent, 29.34 per cent and 20.91 per cent for area, production and yield. Similar findings were also reported by Singh *et al.* (2003). Variation in area under a crop occurs mainly in response to distribution, timeliness and variations in rainfall and other climatic factors, expected prices and availability of crop-specific inputs. All these factors also affect the variations in yield. Further, yield is also affected by outbreak of diseases, pests, and other natural or man-made hazards.

Area, production and yield instability of maize crop in the state of Jammu and Kashmir depicted in Table 4.7 indicated a declining variation in production *i.e.*, from 15.180 per cent to 7.322 per and in yield 14.612 per cent to 11.090 per cent from period I to period II. Implementation of "Accelerated Maize Development Programme" (AMDP) in the State resulted into the adoption of new technology mission which marked decline in instability in production and yield of maize crop, while as the instability in area had increased from 1.850 per cent (period I) to 2.071 per cent (period II). Similar findings were observed by Radha and Prasad (1999).

The results of instability in area, production and yield of maize crop in Jammu region (Table 4.8) revealed that area variability was declined from period I (2.300 per cent) to period II (1.584 per cent). Similarly, for production it was 11.794 per cent and 8.732 per cent and for yield it was 10.641 per cent and 9.330 per cent, respectively during period I and II. The overall instability for area, production and yield was 5.041 per cent, 10.142 per cent and 10.201 per cent, respectively. These results were in close conformity with the results of Atibudhi (2003).

The data from Table 4.9 on coefficient of variation for area, production and yield under maize crop for Doda district, it was 3.941 per cent, 8.260 per cent and 8.891 per cent, respectively during period I and in period II, it was 2.363 per cent, 30.861 per cent and 31.121 per cent, respectively. For overall period instability in area, production and yield was 4.370 per cent, 28.150 per cent and 29.601 per cent, respectively. It was observed that yield had higher instability during the period II and overall period. As far as Udhampur district was concerned the values of coefficient of variation for area was 3.596 per cent and 4.520 per cent for period I and period II, for production it was 9.072 per cent during period I and 17.836 per cent during period II. The variability in yield had increased during the period II (18.207 per cent) as compared to period I (8.645 per cent). It was quite evident that the main contributing factor for inter-year production variability was yield of maize in both the selected districts. These results were in close conformity with the results of Varghese and Rathore (2003) and Raghuwanshi *et al.* (1995). The variability in yield was due to uncertain weather conditions and majority of farmers were using their own farm produced seeds, because due to the high and unaffordable prices of hybrid seeds, lack of technical know-how, non existence of extension services and non-availability of credit facilities, thereby affecting the yield of crop in selected districts.

The instability in area, production and yield of maize crop among selected maize growers of Doda district for the year 2007-08 were analyzed through coefficient of variation (Table 4.10) and registered highest instability in area for medium (56.92 per cent) farmers followed by marginal (46.08 per cent), small (37.18 per cent) and lowest for the large farmers (30.96 per cent) and overall instability in area was 41.65 per cent. Marginal farmers recorded the highest instability of 48.24 per cent in production of maize followed by small farmers (43.41 per cent), medium farmers (39.05 per cent) and large farmers (34.782 per cent), whereas overall instability in production was 44.21 per cent. In yield instability was found to be 43.45 per cent, 31.23 per cent, 47.402 per cent and 29.09 per cent for marginal, small, medium and large farmers, respectively and overall yield instability was 37.74 per cent for sample farmers of Doda district. The results of Table 4.11 for selected maize growers of Udhampur district indicated that instability of 53.82 per cent in area was highest for marginal farmers and lowest instability of 45.03 for large farmers. The variability indices of maize production were found to be higher for medium farmers and also lowest for large farmers, whereas the instability in maize yield was highest in marginal farmers and again lowest for large farmers in the study area. The overall instability was 51.01 per cent, 57.18 per cent and 69.90 per cent for area, production and yield, respectively. However, it was concluded that the instability values for area, production and yield were observed to be lower for large farmers than that of other categories of farmers, indicating the stability attained by large farms under maize crop production. Such type of higher variability of smaller sized maize growers was due to the various socio-economic constraints faced by the farmers was mainly high labour cost, small size of holdings, high cost of micro-nutrients and transportation, non-availability of market for the produce and government was not taking any step for the development of the area.

4.13.3 Decomposition of Maize Production into Area, Yield and Interaction Effects

The data regarding effect of change in area, yield and their interaction (Table 4.12) on the change in production of maize crop between the two periods for India (1987-88 to 1996-97 and 1997-98 to 2006-07), indicated that during period I, the change in area brought about a change in production by 0.714 million tonnes while keeping the productivity constant. Similarly, a change in yield changed the production by 3.892 million tonnes keeping area constant and the interaction effect of area and yield was 2.778 million tonnes on production during the period I. During period II, the change in area and yield brought about a change in production by 2.470 million tonnes and 0.442 million tonnes, respectively. The interaction effect of area and yield was 1.091 million tonnes on production during the period II. But the overall effect of two periods on production was 2.225 million tonnes and 4.225 million tonnes with respect to area and yield, respectively. These results were similar to those of Singh *et al.* (2003). The above discussion clearly shows that during the period I yield of maize showed the better performance in production of maize as compared to change in area but during period II, the area under maize crop had increased production more as compared to the change in yield. However, the overall effect of two periods on production, yield was found to contribute more towards production growth as compared to change in area under maize crop, which attributed to either advancement in technology or higher inputs use.

Table 4.13 indicated that the yield remained the major contributor to the production during period I (0.014 million tonnes) and period II (0.027 million tonnes) as compared to area 0.012 million tonnes and 0.014 million tonnes, respectively for the state of Jammu and Kashmir. The interaction effect of change in area and yield was higher for period II (0.009 million tonnes) as compared to period I (0.005 million tonnes). These findings were supported by Anupama *et al.* (2005) and Chahal and Kataria (2003) but with higher values. In short the yield of maize improved in the state during both the periods but its benefits could not be sustained due to slight increase in area. It might be due to the less availability of quality seeds, fertilizers, pesticides, credit etc. Though the state has provision for the above mentioned inputs but those do not reach the real growers due to lack of extension workers.

As far as Jammu region was concerned, area contributed almost equally and positively in production during the period I (0.018 million tonnes) and period II (0.017 million tonnes), while negative yield effect was recorded during period I (-0.030 million tonnes) and overall period (-0.026 million tonnes), while yield played the significant role in the increase in maize production during the second period. The interaction effect of area and yield had reduced production by (-) 0.001 million tonnes during period I, while as during period II, change in area, yield and interaction effect had a positive effect on production *i.e.*, 0.017 million tonnes, 0.004 million tonnes and 0.001 million tonnes, respectively. For the overall period the area effect was 0.056 million tonnes on production, area and yield interactive effect had decreased production by (-) 0.004

million tonnes. These findings were supported by Chahal and Kataria (2003). Thus, it may be attributed to change in area, which had affected maize production positively in Jammu region of J&K State and change in yield had negatively affected production because in the traditional maize growing areas, most farmers still grow local maize varieties during the rainy season and seed replacement is very low.

Area effect, yield effect and their interaction effect on the change in production of maize crop (Table 4.15) showed that area positively contributed to the maize production in selected districts (Doda and Udhampur) during all the periods. It was 0.008 million tonnes and 0.001 million tonnes in period I, 0.004 million tonnes and 0.012 million tonnes in period II and 0.010 million tonnes and 0.027 million tonnes during the overall period for Doda and Udhampur districts, respectively. The contribution of yield effect and interaction effect to the maize production dropped considerably and reduced production in Doda district by (-) 0.013 million tonnes and (-) 0.001 million tonnes during period I and (-) 0.024 million tonnes and (-) 0.001 million tonnes in period II and in Udhampur district also, the yield effect was negative and reduced production by (-) 0.016 million tonnes during period II but positive for period I (0.009 million tonnes). Whereas the interaction effect of area and yield was positive for the period I, II and negative for overall period. Varghese and Rathore (2003) were of similar opinion. The reasons for negative effect on yield was that the most of the *kharif* maize is cultivated in rainfed conditions and maize cultivation operations were performed traditionally mode with less number of irrigations, most of the farmers were growing other crops like paddy and vegetables under the irrigated conditions. Further, yield decreased largely because the farmers increasingly cultivated traditional cultivars and did not make the use of inorganic fertilizers, pesticides and other chemicals which are used for the control of weed. They lacked technical and scientific know-how and were facing the problems like non-availability of hybrid seeds and fertilizers at the reasonable rates. Weeding and other crop care/management practices which are considered to be the most important factors affecting any crop production were lacking in the sample area.

4.13.4 Adoption of New Maize Technology

The results on adoption index calculated for individual farmers (Table

4.16), included adoption of improved variety, adoption of recommended doses of chemical fertilizers, recommended number of irrigation etc showed that 49 per cent of the farmers fell under the category of low adoption (less than 33 per cent adoption index), 40 per cent belonged to the category of medium adoption (33-66 per cent adoption index) and only 11 per cent belonged to high adoption (more than 66 per cent adoption index). Kumar *et al.* (2004), Chandra and Singh (1992) and Anupama *et al.* (2005) also observed the similar findings. The main reason for the low level of adoption was that the poor farmers were unable to purchase the required quantity of fertilizers, improved seed to be applied for the maize crop. Most of the farmers also complained that government was not taking any step for providing these inputs on subsidized rates. Moreover non-availability of water for irrigating the fields with recommended times thus, resulted in rainfed maize crop.

4.13.5 Cultivar-wise Area under Maize

Out of the total area under maize crop, marginal, small, medium and large farmers were allocating maximum area under traditional varieties as compared to hybrid varieties in the sample area. Around 15.48 per cent, 19.67 per cent, 43.44 per cent and 32.65 per cent, respectively area under maize was occupied by hybrid cultivars by marginal, small, medium and large farmers and 84.52 per cent, 80.33 per cent, 56.56 per cent and 67.35 per cent by traditional cultivars. On an average 21.78 per cent area was under hybrid varieties and 78.22 per cent under traditional varieties among the sample farms. This revealed that majority of farmers were cultivating the traditional varieties. This was probably due to unavailability of hybrid seeds, high and unaffordable prices for hybrid seeds, education level of the household, existence of extension services and access to credit facilities, thereby affecting the adoption level of farmer. Badal and Singh (2001) were of the same opinion.

4.13.6 Sources of Seed Material of Maize Crop

The results on sources of seed materials of maize crop (Table 4.18) showed that across the entire farm category, majority of the farmers were using their own farm produced seed, some purchased from dealers, others from other farmers and the farmers who purchased from government agencies had negligible number. It was unfortunate that not a single farmer changed the seed in every crop season. Seed dealers sold seed at a very high price which every maize grower could not afford to buy and depended upon using home produced of poor quality seeds which ultimately resulted into technical inefficiency and less production. These findings are supported by Anupama *et al.* (2005).

4.13.7 Comparative Economics and Impact of Improved Maize Technology over Traditional Maize Variety

Cost of cultivation for hybrid cultivars was highest (₹6748.63/ha) as compared to traditional cultivars (₹4685.60/ha), but the net returns were more for hybrid cultivars than traditional one *i.e.*, ₹8737.37/ha and ₹5472.40/ha, respectively. The per hectare output value was ₹15486.00 for hybrid and ₹10158.00 for traditional cultivars and the output-input ratio of hybrid maize (2.29) was also greater than traditional (2.16) cultivars. This discerns that the improved cultivars had increased the yield. No doubt, the cost of hybrid variety was more than traditional variety, but the returns were large enough to cover the increased cost leaving more returns than traditional variety. There was increase in cost of cultivation with the use of improved cultivars due to additional expenditures on seeds, fertilizers, irrigation and other inputs. However, the input cost of production was low for the traditional varieties as compared to the hybrid varieties, but the net returns was more for hybrids, hence proved economic viability of hybrid cultivars. Hybrid maize can also be preferred not only due to better returns but also due to the less requirement of water as compared to other competitive crops like paddy in the study area. Further, the impact of improved maize technology over traditional maize on yield as well as on the unit cost of production indicated that maize yield had increased from 16.93 qtls/ha (traditional) to 25.81qtls/ha with the use of hybrid cultivar and the unit cost of production had reduced to ₹252.32 for hybrid from ₹338.52 for the traditional variety. The rationale behind improvement lies in the enhancement of crop productivity and reduction in per unit cost of production. The improved maize cultivar satisfied these criteria. In other words, the cost reduced by 25.46 per cent for hybrid varieties when compared to traditional varieties. In nutshell, the improved cultivars were able to increase the crop yield and to reduce the cost of production significantly, hence the production technology had improved. These findings are supported by Anupama *et al.* (2005).

4.13.8 Socio-economic Constraints to Maize Production by Farmers

The results on Socio-economic constraints faced (Table 4.20) by the farmers in adoption of improved maize technology was high labour cost, small size of holdings, high cost of micro-nutrients, pesticides and transportation, non-availability of market, absence of technical knowledge, educated members go outside, labour not available when needed and government was not taking any step for the development of the study area. The main constraint was that the adoption of improved maize technology was expensive and hence farmers had difficulty in adopting it, because the hybrid seeds, fertilizers, pesticides and micro-nutrients are more expensive and unavailable which contributed to the additional cost of cultivation and government is not providing these inputs timely. Due to either lack or high cost of transportation, the crop could not reach the right market at the right time, which compelled the growers to sell the crop at throw away prices to local agents, thus resulting into their disinterest in adopting modern technology of maize crop cultivation. Under such circumstances, the government organization has greater responsibility to make available the quality seed material to the resource poor farmers at affordable rates. Furthermore, there is also need to assure the farmers about the marketability of their increased produce, as road networks in the maize growing regions of sample area are not well developed. Markets for food grains in general and maize in particular, are very thinly spread throughout the maize growing regions. Most maize production is sold in local village markets. The market margin for maize growers was 25 per cent. Farmers continue to sell their produce in the local village market because: (1) when grains are sold outside the village, transportation costs tend to be higher than marginal returns due to price difference, and (2) farmers tend to sell to local traders, especially if they need to pay back any credit they may have taken out to purchase inputs and for consumption purposes. Farmers were of the opinion that there is no other reliable way to sell their produce, as the volume is often very low. The other constraint in crop production was the non-availability of labour and lack of technical knowledge which severely limit actual production in spite of great inherent production. This could probably be explained by the fact that very high education (University and college) attenuates the desire for farming and therefore, the farmer probably concentrates on salaried employment instead of farming. But it was concluded that farmers with more education respond more readily to new technology and produces closer to the frontier output. The finding is also consistent with results on structural adjustment and economic efficiency of rice farmers in Northern Ghana by Awudu and Huffman (2000). In addition to this, the farmers who were less educated do not attend the agricultural extension training which ultimately results into lack of technical knowledge. Small size of holding is also the constraint in crop production which is supported by the findings of Shanmugam and Venkataramani (2006), in which they concluded that large size holdings helps to improve the efficiency of farmer in crop production.

4.14 Economic Efficiency

4.14.1 Resource-use Efficiency

Ordinary least square estimates of Cobb-Douglas production function showed (Table 4.21) the co-efficient of multiple determinations (R^2) indicated that 44 per cent of the variation in the yield could be explained by the variables considered in the model for the maize crop. Similar findings were observed by Anupama *et al.* (2005). The average performance of the sample farmers and the estimated coefficient of labour was positive (0.338) and statistically significant at one per cent level of significance, which indicated that maize yield could be increased 0.338 per cent by using more human labour, while keeping other factors constant. These results are in consistency with the findings of Hasan (2008) and Mohiuddin *et al.* (2007). The statistical significance (at 1 per cent level of significance) for the regression coefficient of capital implied that 0.306 per cent production could be increased with one per cent increase in capital used. Similar findings were observed by Anupama *et al.* (2005). The results further indicated that the coefficients of N (0.324) and K (0.202) were positive and significant at one per cent level of significance, indicating there by that maize yield could be increased by 0.324 and 0.202 per cent by employing more units of these inputs. It is in conformity with the findings of Hasan (2008) and Oluwatayo *et al.* (2008). Phosphorus (P) with statistically significant negative value (-) 0.097 revealed the excessive use of P and if phosphorus was increased furthermore, it could rather reduce the maize production by 0.097 per cent. Chowdhury (1992) found a negative and insignificant coefficient for fertilizer in maize production. In the case of irrigation, the co-efficient were positive (0.043) but statistically non- significant. This implied that irrigation had not much more role to play in the production of maize crop in the study area.

4.14.2 Allocative Efficiency of Sample Farms

The MVP of various inputs namely labour, capital, irrigation and fertilizers (N,P,K) were worked out at their geometric mean level in order to compare the estimates of allocative efficiency and their productivity. The estimated MVP and MFC ratio *i.e.* allocative efficiency of different inputs in the study area could be seen from Table 4.22. In case of irrigation and fertilizers (N,K) the ratio of MVP and MFC were greater than one and positive, which indicated the increasing returns to scale. Therefore, it means that the maize farmers were utilizing these resources sub-optimally and if amount of these resources were increased, they could increase the output more than proportionally, which is in conformity to the findings of Duloy (1959) and Hasan (2008). In case of capital and human labour used, though their contribution to maize output was positive which was clear from their positive regression coefficient, but was less than one, indicating thereby diminishing returns to scale which meant that one per cent increase in these inputs could increase the maize output less than one per cent or the total output could increase at a decreasing rate. The ratio of fertilizer (P) was negative but greater than one hence, this input for maize production was not used efficiently in sample farms, which lead to the reduction of yield. These findings were supported by Mohiuddin *et al.* (2007).

4.14.3 Technical Efficiency of Sample Farms

The estimates of stochastic frontier showed the best practice *i.e.*, efficient use of available technology. The information from Table 4.23 indicated that the dependent variable included in the model was output of maize crop. The independent variables included human labour, capital, fertilizers (N, P, K) and irrigation. The regression coefficient of labour, capital, and fertilizers (N,K) were positive and significant at one per cent level of significance, indicating 0.378 per cent, 0.336 per cent, 0.244 per cent and 0.292 per cent increase in maize production with one per cent increase in these inputs. The value of regression coefficient of irrigation was positive but insignificant. The positive sign implies that one per cent increase in irrigation, keeping other factors constant, could increase the yield by 0.225 per cent and the regression coefficient of fertilizer (P) was negative and insignificant, the negative algebraic sign and statistical insignificance for coefficient Phosphorus was due to its over use. The estimated value of λ was 4.219 and σ was 0.455, which were significantly different from zero indicating a good fit and the correctness of the distributional assumptions specified. The value of λ was more than one, employing the dominance of one sided component U_i in E_i and thus indicated high degree of technical inefficiency. In other words the inefficiency component was not dominated by the random factors. The variance ratio γ showed that the farm specific variability contributed more to the variation in yield, which means that variation in output from frontier is attributed to technical inefficiency. The value of γ was 0.946 meaning thereby that about 94 per cent of the differences between the observed and the maximum production frontier outputs were due to the factors, which were under farmers' control. The stochastic frontier analysis further showed that 94 per cent of the observed inefficiency was due to farmers' inefficiency in decision making and only 6 per cent of it was due to random factors outside their control. These results are in conformity with Shanmugam and Venkataramani (2006) and Ogundari, (2006).

The estimated value of σ^2_v and σ^2u were 0.011 and 0.196 respectively, which means that the differences between the observed and frontier output were due to inefficiency and not due to chance alone. Since the frontier parameters indicated the maximum possible contribution of each input to output, when the inputs are utilized efficiently with the help of best farming techniques.

The results on elasticity of production and return to scale could be seen from the Table 4.24. The regression coefficients in the frontier production function are the production elasticities, and their sum indicates the returns to scale. All of the individual elasticities of production were less than one, but the summation of elasticities of different inputs were greater than one (1.092>1) indicating increasing returns to scale. This indicated that farmers in general, allocated their resources in the irrational zone of production (stage 1) where the increasing returns to scale prevails, that is, if all the inputs specified in the function were increased by one per cent, output would have increased by 1.092 per cent, which also means that there is the area potentiality of increasing maize production further till the completion of second stage of production (as among the three stages of "law of variable proportions" second stage of production function is rational zone of production) These results are in conformity with the findings of Dolisca and Jolly (2008), Adeleke *et al.* (2008) and Mohiuddin *et al.* (2007). Labour appeared to be the most important factor of production with highest elasticity of 0.378, among all other factors showing the labour-intensive nature of farming in the study area. Statistically significant elasticity suggested that 1 per cent increase in labour would result in an increase of 0.378 per cent in production of maize crop. These results are in agreement with the work reported by Amaza and Olayemi (2000).

4.14.4 Efficiency Level of Sample Farms

Technical efficiency (Table 4.25) indices range from minimum of 8 per cent to maximum of 96 per cent for the farmers in the sample area, with an average of 52 per cent indicated on an average, the sample farms operated 48 per cent below the frontier output levels. This means that the maize output of the average farmer could be increased by 48 per cent by adopting the technology followed by the best practice farmers. The level of technical efficiency observed in this study appeared to be lower than the 58-59 per cent efficiency reported by Bravo-Ureta and Evesan (1994) for farmers in eastern Peru, Anupama *et al.* (2005) among the maize producers in Madhya Pradesh (77 per cent), Audibert (1997) among paddy farmers in Mali (70 per cent) and Wadud and White (2000) for rice farmers in Bangladesh (75 per cent). The results are also in conformity with the results of other studies that have shown the existence of substantial technical inefficiencies in developing agricultural economies. For example, high technical inefficiency has been found to exist among paddy growers in Jammu region of J&K State (63 per cent) by Kachroo *et al.* (2008). The indices in Table 4.25 show that the technical efficiency of the sampled farmers was less than 1 (less than 100 per cent) implying that all the farmers in the study area were producing below the maximum efficiency of frontier. Moreover, farmers who got the highest score of technical efficiency above 90 per cent were 9.16 per cent. On the contrary, the group of the lowest score of technical efficiency under 40 per cent was 23.75 per cent. From this estimation, maximum technical efficiency has not yet achieved probably because most of the farmers produced maize crop with the use of a given mix of production inputs, inefficient tools, traditional seed etc. Level of experience, age and low level of formal education were the major factors that culminated to influence the magnitude of the farmers' technical efficiency.

4.14.5 Estimated Potential Yield

Under different efficiency levels the estimated average potential yield (Table 4.26) for the farmers was 18.36 qtls/ha and average actual yield was 10.82 qtls/ha, this could go up to 25.64 qtls/ha among the most efficient farms having the technical efficiency level above 90 per cent. However, there was a range of actual yield between 7.53 qtls/ha and 14.88 qtls/ha (technically most inefficient and technically most efficient farmers, respectively). This showed that there was intra-variation in yield among farmers in the same region. The farmers having efficiency level less than 40 per cent showed the actual yield 7.53 qtls/ha, 9.31 qtls/ha in the efficiency level of 40-50 per cent, 9.46 qtls/ha in 50-60, 10.66 qtls/ha in 60-70, 11.02 qtls/ha in 70-80 and 12.90 qtls/ha for the farmers under the efficiency level of 80-90 per cent. The potential yield showed that the actual yield could go up to 13.98 qtls/ha (less than 40 range), 15.09 qtls/ha (40-50 range), 15.70 qtls/ha (50-60 range), 17.47 qtls/ha (60-70 range), 18.57 qtls/ha (70-80 range), 22.12 qtls/ha (80-90 range) and 25.64 qtls/ha (above 90 range), if the farmers were technically most efficient or used the resources properly. Thus, the difference between the potential and actual yield was 6.45 qtls/ha, 5.78 qtls/ha, 6.24 qtls/ha, 6.81 qtls/ha, 7.55 qtls/ha and 9.22 qtls/ha, respectively in above mentioned efficiency levels. These results are in conformity with Kibaara, (2005) and Ingosi, (2005).

4.14.6 Input Use and Technical Efficiency

The details about input use across various levels of technical efficiency (Table 4.27) showed that technically the most efficient producers (greater than 90 per cent efficiency level) had the highest average yield of 14.88 qtls/ha and used the following combination of inputs: 46.44 Kg/ha, 11.56

Kg/ha, 12.20 Kg/ha (N,P,K), 32.29 Kg/ha seed and 40.78 mandays/ha. However, though efficient among the sampled maize growers, they still used inputs below the recommended rates. For example, the amount of fertilizer used by the most efficient farmer was much below the recommended rate of 60 Kg/ha, 40 Kg/ha, 20 Kg/ha (N,P,K), 35-40 Kg/ha of seed and about 83 mandays days/ha. In comparison, the least efficient producers (less than 40 per cent) who had the yield of merely 7.53 qtls/ha, were using the input far-far less than the recommended dosages (23.08 Kg/ha, 7.01 Kg/ha, 4.79 Kg/ha (N,P,K), 20.58 Kg/ha and 50 mandays/ha). The labor used by this cluster is greater than that of the most efficient producers, analysis showed that inefficient cluster source higher proportion of labour from family as compared to the most efficient group, family labor is therefore cheaper than fertilizer, indication of input substitution among the least technically efficient producers. Similar findings were observed by Kibaara, (2005) for the technical efficiency in Kenyan's maize production.

4.14.7 Factors Affecting Technical Efficiency

The multiple regression results (Table 4.28) concluded that there is interaction between the level of farm efficiency and socio-economic variables. Interpretation of results yields a number of additional insights into the process of factors affecting technical efficiency among maize growers in Jammu region of J&K State. The coefficient for constant is positive and highly significant, thereby indicated that per hectare production was high with the given mean quantities of inputs. The dummy variable for age was negative (-0.001) and insignificant, suggesting that younger farmers, who were less than 50 years, were more efficient than the older ones. This means that the aged farmers were not adopting the new technology for maize crop production in the study area. These results are consistent with the findings of Bravo-Ureta and Evenson (1994) and Abdulai and Eberlin (2001). Hussain (1989) also observed that older farmers are less likely to have contacts with extension workers and are less willing to adopt new practices and modern inputs. The reason for this is probably because the age variable picks up the effects of physical strength as well as farming experience of the household head. Although farmers become more skillful as they grow older, the learning by doing effect is attenuated as they approach middle age, as their physical strength starts to decline (Liu and Zhung, 2000). Similar conclusions were made by Awudu and Huffman (2000). The variable education (0.023) was positive and significant at 1 per cent level of probability, had positive impacts on the technical efficiency level for maize farmers. This implies that farmers with better education were technically more efficient. These finding are similar to Dey *et al.* (2000) and Liu and Zhuang (2000), who found that farm efficiency increases with level of education. Increased level of educational achievement may lead to a better evaluation of the importance of better farming, decision making, including the efficient use of inputs. A good level of education enhances a farmer's ability to seek, interpret and make good use of information and production inputs. A study by Seyoum *et al.* (2000) on technical efficiency and productivity of maize producers in Eastern Ethiopia concluded that farmers with more education respond more readily to new technology and produce closer to the frontier output. This finding is also consistent with results on structural adjustment and economic efficiency of rice farmers in Northern Ghana by Awudu and Huffman (2000). From these results, it can be concluded that a maize producer need education to know the correct amount of fertilizer to be applied, correct seed rate and general management of the farm. The positive and statistically significant coefficients for the proportion of female workers under study indicated that females were contributing a lot in the various operations of maize cultivation e.g, in harvesting, storing of produce, weeding etc, their participation was more than their male counterparts. The reason for that their male members mostly migrated to other places in search of other jobs either as labourers or as employees. Thus, it could be seen that female workers had better opportunities to carry out frequent follow-up and supervision of the farm activities on their plots. This is consistent with the findings of Onyenweaku and Effiong, (2005) and Dolisca and Jolly (2008). The variable for the proportion of children in the family was positive, but statistically insignificant. Thus indicated that children had nothing to do with efficiency *i.e.*, they were neither technically efficient nor inefficient. The farm size showed a significant and positive relationship with regard to technical efficiency, which means that the farm size had a major influence on the efficiency of sample farm. Similar conclusions were made by Kumar *et al.* (2005).

4.15 Economics (Costs and Returns) on Maize Production

4.15.1 Cost Structure for Maize Cultivation in Doda District

The results of cost structure for maize cultivation (Table 4.29) indicated that per hectare total cost of maize crop cultivation was highest for marginal farmers (₹13287.60/ha) followed by medium farmers (₹13187.24/ha), small farmers (₹12863.35/ha) and lowest for large farmers (₹12792.22/ha) with overall average cost of ₹12882.83/ha. The cost for large farmers was lower due to lower amount of inputs used for maize cultivation in study area. Similar findings were reported by Singh *et al.* (2007). Among the different inputs items used in operational cost, overall average cost of human labour (₹4193.13/ha) was major component followed by fertilizers and manures (₹1057.02/ha), seed (₹747.43/ha), bullock labour (₹252.84/ha), interest on working capital (₹244.22/ha) and miscellaneous expenditures (₹213.50/ha), which contributed 32.54 per cent, 8.20 per cent, 5.80 per cent, 1.96 per cent, 1.89 per cent and 1.65 per cent, respectively to the total cost. It was observed that the cost of human labour was highest for marginal farmers (₹4311.96/ha). This was due to higher family labour utilization in marginal size farms than other category of farmers, among the fixed cost overall average rental value of owned land accounted for 35.56 per cent followed by interest on fixed capital (10.97 per cent) and 1.38 per cent of the total cost of maize cultivation for depreciation on implements and farms buildings and similar findings were observed by Neeraj *et al.* (2003) and Kumar and Kumar (2004).

4.15.2 Returns from Maize Production in Doda District

The returns from maize production in Doda District (Table 4.30) indicated that the gross income from maize crop was highest for marginal farmers (₹18009.00/ha) followed by small (₹17712.00/ha), medium (₹17352.00/ha) and large farmers (₹17118/ha), respectively. The overall average gross income per hectare was worked out to be ₹17500.27/ha. On an average the total cost on A_1 and A_2, B_1, B_2, C_1, C_2 and cost C_3 were worked out at ₹8476.34/ha, ₹8683.36/ha, ₹8937.98/ha, ₹12645.90/ha, ₹12882.83/ha and ₹14171.11/ha, respectively. All costs increased with the decrease of farm size for maize crop, all the costs were higher in case of marginal farmers than large farmers. The reason attributed is on higher investment of input on marginal farmers, because in intensive farming marginal farmers were used more family labour, farm produced seed, FYM and they were able to make more use of inputs as compared to large farmers. It was further observed that on overall average return over each rupee invested in maize production were highest in case of A_1 and A_2 (2.09) followed by cost B_1 (2.04), B_2 (1.98), C_1 (1.40), C_2 (1.35) and C_3 (1.23) and was more than one for all costs, which suggested that the cultivation of maize is economically profitable and each one rupee spent in maize cultivation would yield return of ₹2.09 in case of A_1 and A_2, ₹2.04 (B_1), ₹1.98 (B_2), ₹1.40 (C_1), ₹1.35 (C_2) and ₹1.23 for C3, respectively. Similar findings were observed by Kumar and Kumar (2004) and Neeraj *et al.* (2003).

4.15.3 Cost Structure for Maize Cultivation in Udhampur District

The results of cost structure for maize cultivation (Table 4.31) indicated that per hectare total cost of maize crop cultivation was highest for marginal farmers (₹10577.76/ha) followed by medium farmers (₹10465.12/ha), small farmers (₹10202.08/ha) and lowest for large farmers (₹10178.73/ha). The cost for large farmers was lower due to lower amount of inputs used for maize cultivation in Udhampur district was similar to Doda district. Similar findings were given by Singh *et al.* (2007). Among the different inputs items used in operational cost, overall average cost of human labour (₹3818.89/ha) was major component followed by fertilizers and manures (₹1332.48/ha), seed (₹384.33/ha), interest on working capital (₹259.53/ha) bullock labour (₹234.41/ha), and miscellaneous expenditures (₹190.21/ha), which contributed 36.61 per cent, 12.77 per cent, 3.68 per cent, 2.48 per cent, 2.24 per cent and 1.82 per cent, respectively to the total cost. It was observed that the cost of human labour was highest for marginal farmers (₹3908.08/ha) followed by small farmers (₹3815.33/ha), medium farmers (₹3708.18/ha) and large farmers (₹3673.28/ha). This trend was due to higher family utilization in marginal size farms than large farmers, among the fixed cost overall average rental value of owned land accounted for 29.06 per cent followed by interest on fixed capital (9.48 per cent) and 1.82 per cent of the total cost of maize cultivation for depreciation on implements and farms buildings and similar findings were given by Neeraj *et al.* (2003) and Kumar and Kumar (2004).

4.15.4 Returns from Maize Production in Udhampur District

The returns from maize production in Udhampur District (Table 4.32) indicated that the gross income from maize crop was highest for marginal farmers (₹15282.00/ha) followed by small (₹13689.00/ha), medium (₹13383.00/ha) and large farmers (₹13212.00/ha), respectively. The overall average per hectare gross income was worked out to be ₹14276.85/ha. On an average the total cost on A_1 and $A_{2,}$ $B_{1,}$ $B_{2,}$ $C_{1,}$ C_2 and cost C_3 were worked out at ₹7008.35/ha, ₹7417.51/ha, ₹8448.54/ha, ₹9209.98/ha, 10430.34/ha and ₹11473.92/ha, respectively. All costs increased with the decrease of farm size for maize crop and were higher in case of marginal farmers than large farmers. This was due to higher investment of inputs by marginal farmers, because in intensive farming farmers with small size of holdings used more family labour, farm produced seed, FYM and were able to make more use of inputs as compared to large farmers. It was further observed that on overall average return over each rupee invested in maize production were highest in case of A_1 and A_2 (2.03) followed by cost B_1 (1.92), B_2 (1.68), C_1 (1.55), C_2 (1.36) and C_3 (1.24) and was more than one for all costs, which suggested that the cultivation of maize was economically profitable and each one rupee spent in maize cultivation would yield return of ₹2.03 in case of A_1 and A_2, ₹1.92 (B_1), ₹1.68 (B_2), ₹1.55 (C_1), ₹1.36 (C_2) and ₹1.24 for C_3, respectively. Similar findings were observed by Kumar and Kumar (2004) and Neeraj *et al.* (2003).

Chapter 5
Summary and Conclusion

This chapter explains the summary and conclusion derived from the present study on "Economic efficiency of maize production in Jammu region of J&K state" during the year 2008-09. Udhampur and Doda districts of Jammu region were selected because these two areas covered the maximum area under maize cultivation. Three blocks from each district were selected as secondary stage units. From each block two villages were selected as the third stage units and the ultimate units were twenty households from each village so as to constitute sample size of 240 households. Primary data required for working out resource use efficiency, technical efficiency and factors affecting on technical efficiency were collected by survey method with the help of specially designed schedules for the purpose by interviewing the farmers personally and for calculating compound growth rates and instability for area, production and yield of maize, secondary data was collected from the various published sources such as bulletins of the ministry of agriculture, Govt. of India, Directorate of Economics and Statistics, Jammu and Kashmir Government.

The data collected were subjected to analysis for examining the objectives of the investigation *viz.* estimation of growth rates, instability, adoption index and production function for the estimation of resource use efficiency, technical efficiency and factors affecting technical efficiency and economics of maize production. A stochastic frontier approach was used to generate technical efficiency estimates using Limdep (Green, 2002). Compound growth rates and coefficient of variation were worked out to know the performance of maize production and instability, analyzing effect of area, productivity and their interaction in increasing the maize production was examined by using differential equation and various cost concepts were employed to explain the economics of maize production. The salient findings of the present investigation had been summarized in this chapter.

5.1 Production Status of Maize

5.1.1 Growth Performance of Maize Production

At national level the growth rate of maize in terms of area, production and yield categorized in different periods (Period I 1987-97, Period II 1997-07) showed positive trend in area, production and yield in all the periods.

At state level the overall growth rate of 2 decades data revealed that area, production and yield showed significantly positive trend. However, during the period-II (1997-07) the area under maize had increased, whereas the yield and production recorded significantly negative growth.

At Jammu region of J&K State, it was observed that area in maize had increased in all the periods, where as yield and production of maize recorded negative growth rate during period II (1997-07), which was contrary to period I (1987-97), where both yield and production was found significantly positive.

The growth performance of maize crop in sample districts of Jammu region revealed that the growth rate for area in Doda district was positive during all the periods, where as the production and yield showed the negative trends in all the periods. Udhampur district showed the negative growth trend in area for the period I and positive for the period II and overall period in case of area, whereas the growth in production and yield was positive in the period I and negative in period II and overall period.

5.1.2 Instability in Maize Crop Cultivation

At national level the instability of maize crop in terms of area, production and yield vary to a large extent *i.e.*, 9.55 per cent in area,

29.34 per cent in production and 20.9 per cent in yield. However in case of different periods the instability in area of maize during period I and II is meager whereas, the yield and production varies from 13.38 per cent to 13.35 per cent.

At the state level the similar trend in terms of area, production and yield has been observed, which varies from 4.52 per cent to 11.3 per cent and indicates that the instability factor in J&K State is slightly lesser than national level. The instability during period II for the area increased (2.07 per cent) but decreased in case of production and yield than period I.

At Jammu region of state the instability in area, yield and production of maize was 5.04 per cent, 10.20 per cent, 10.14 per cent, respectively. However, the instability in terms of area, production and yield of maize was lesser than that of period I.

Among sample districts, the highest instability in area was observed in Udhampur district and that of production and yield was observed in Doda district. Both Doda and Udhampur district of Jammu observed higher instability in area, production and yield under period II than period I.

The instability under the sample farmers of Doda district revealed that overall instability in terms of area, production and yield was 41.6 per cent, 44.21 per cent and 37.73 per cent, respectively. The instability in terms of land holding of the farmer under area was in order of medium (56.92 per cent) > marginal (46.08 per cent) > small (37.18 per cent) > large (30.96 per cent). Whereas, in case of production highest instability was observed in marginal land holdings (48.2 per cent) followed by 43.4 per cent in small holdings, 39.05 per cent in medium holdings and 34.78 per cent under large holdings. In case of yield of maize the minimum instability was observed in large land holdings (29.09 per cent) followed by small holdings (31.22 per cent), marginal (43.45 per cent) and medium (47.40 per cent).

The sample farmers at Udhampur district under different land holdings showed that the instability in case of area, production and yield of maize was higher than that of Doda district. Large holding farmers showed lesser instability of area under maize over medium (49.58 per cent), small (51.58 per cent) and marginal (53.82 per cent). Whereas in case of production the highest instability was observed in medium land holdings (63.24 per cent) followed by small (63.21 per cent), marginal (54.05 per cent) and large (47.31 per cent). However, instability in terms of maize yield under different holdings showed that least instability was observed under large farmers (54.48 per cent) followed by medium farmers (67.42 per cent), small farmers (67.49 per cent) and that of marginal (76.65 per cent).

5.1.3 Decomposition of Maize Production into Area, Yield and Interaction Effects

The results regarding to the effect of change in area, yield and their interaction on the change in production of maize crop between the two periods for India showed that during the period I yield of maize had the better performance in production as compared to change in area but during period II, the area under maize crop had increased production as compared to the change in yield and the interaction effect of area and yield in both the periods helped in maize crop production during both the periods. These positive effects of both area and yield on production was due to the significant shift occurred during nineties in maize production as the winter maize cultivation expanded rapidly in major maize growing states of India. The new paradigm shift in the seed and price policy of cereals enhanced support to the export of cereals, which would bring significant expansion in maize area.

The decomposition analysis for Jammu and Kashmir State showed that the yield remained the major contributor to the increased maize production during period I and II as compared to area in both the periods. The Interactional effect of area and yield was higher in period II as compared to period I. The positive yield effect on production showed that there is potential to increase the production of maize by bringing more area under high yielding varieties, increasing the application of purchased inputs could be made for enhancing the production of crop.

Jammu region showed that area had positive effect on production during period I and II and had negative effect of yield on production during period I and overall period, but played the positive role during period II. The interaction effect of area and yield had reduced the production during period I, but increased the production during the period

II. It was further observed that the area remained a major contributor to the production during both the periods in the region over a period of time, while as the negative yield effects dominated the area effects in period I and overall period, which could be reduced either by advancement in technology or higher inputs used.

The decomposition of maize production into area, yield and interaction effects for selected districts (Doda and Udhampur) showed that area contributed significantly to the increase in maize production during all the periods, where as the yield contributed negatively during both the period in Doda and Udhampur districts. Yield increased production during period I and reduced during period II. The interaction effect was positive for period I and II in Udhampur district and negative for Doda district. It was observed that there was almost negative effect of yield on production for the selected districts. There were so many socio-economic constraints which decreased the yield in sample area and farmers mostly cultivated maize under rainfed conditions with less number of irrigation. Constraints and causes should further be studied because minimization of constraints will help in increasing the yield level and production in the sample area.

5.1.4 Adoption of New Maize Technology

It was observed from the results that a major portion (49 per cent) of farmers for the study area was low adopters (less than 33 per cent of adoption index) and 40 per cent belonged to the category of medium adoption (33-66 per cent of adoption index). A small fraction (nearly11 per cent) of farmers were coming under the category of high level (more than 66 per cent adoption index) of adoption. It proved that the adoption of new technology had not made an appreciable headway and traditional methods of crop cultivation were still prominent on sample farms.

5.1.5 Cultivar-wise Area under Maize

Out of the total area under maize crop, marginal, small, medium and large farmers were allocating maximum area under traditional varieties as compared to hybrid varieties in the sample area. This is probably because of high prices for hybrid seeds, making them unaffordable to most subsistence maize producers, education level of the household, existence of extension services and access to credit facilities, easily and timely availability of traditional seed, because it produced on the same field.Therefore, government should take step to provide the hybrid maize on subsidies to encourage the farmers for cultivating the hybrid varieties so that the production could be increased.

5.1.6 Sources of Seed Material of Maize Crop

The study revealed that majority of the farmers depended on the farm produced seed and thereafter purchase the seed from the dealers.

5.1.7 Economics and Impact of Improved Maize Technology over Traditional Maize Variety

The cost of cultivation of traditional and hybrid cultivars revealed that there was significant increase in the cost of cultivation while using improved cultivars due to its additional requirement of fertilizers, irrigation, plant protection chemicals when compared to the traditional varieties. However, the significant increase in yield, the unit cost of production was much lower in case of improved cultivars. The impact of improved maize over traditional ones showed increase in yield and reduction in unit cost of production. Therefore, the farmers can very well adopt modern technology by replacing the traditional ones and get better returns.

5.1.8 Socio-economic Constraints to Maize Production by Farmers

The farmers faced many socio-economic constraints. Among those were high expenditure in the inputs mainly hybrid seeds, fertilizers, pesticides and micro-nutrients, which contributed to the additional cost of cultivation and government not providing these inputs on subsidies followed by non-remunerative price of the maize crop. Besides this, the high cost of transportation due to non-availability of markets in the locality forces the farmers to sell their product at low prices to local agents and thereby earn less profit, which restricts them to use the improved maize technology. Thus government organization has greater responsibility to make available the quality seed material to the resource poor farmers at affordable rates and proper government agency to procure the produce was also a serious constraint faced by the farmers.

5.2 Resource-use Efficiency

The value of co-efficient of multiple determinations (R^2) indicated that 44 per cent of the variation in the yield could be explained by the variables considered in the model for the maize crop. The value of regression coefficient of human labour, capital used and fertilizers (N and K) were positive indicating that statistically significant and positive values of estimated coefficients could increase per hectare yield by employing more units of these inputs in producing the maize crop and these inputs were productive inputs for successive production of maize crop in selected areas of Jammu region. The value of the production coefficient for fertilizer (P) was negative and significant. The negative algebraic sign for P might be due to it's over use.

5.2.1 Allocative Efficiency of Sample Farms

The allocative efficiency of farmers *i.e.* ratio of MVP and MFC, showed that the value of inputs like irrigation and fertilizers (N,K) were greater than one and positive that means the maize farmers did not avail the opportunity of using the optimal amounts of those inputs. The main reason may be due to high cost of fertilizers and non-availability of irrigation facility in that area. The farmers could increase the output per hectare by making more use of these inputs so that more profit can be obtained. This is an indication that the maize producer can still benefit by increasing fertilizer and irrigation use. The allocative efficiency for capital and human labour used showed that the use of these inputs had gone beyond the economic optima, which is due to more dependence of farmers on agriculture. If the farmers use these inputs efficiently it might help them in increasing the output within given resources. The ratio of fertilizer (P), was negative but greater than one, means that this input for maize production was not used efficiently in sample farms, thereby lead to the reduction of yield. The economic efficiency of the maize growers in sample districts can be improved by providing good quality seed materials of improved maize cultivars and easy credit supply so as to purchase chemical fertilizers.

5.2.2 Technical Efficiency of Sample Farms

The estimates of stochastic frontier showed that the regression coefficient of labour, capital, and fertilizers (N,K) were positive and significant at one per cent level of significance, indicating that increase in maize production with one per cent increase in these inputs. The positive value for irrigation implies one per cent increase in irrigation also increased the yield of maize but the negative value for fertilizer (P) indicated that this fertilizer was not used efficiently hence leads to reduction in yield. The value of λ was 0.94 that about 94 per cent of the differences between the observed and the maximum production frontier outputs were due to the factors, which were under farmer's control and farm specific variability contributed more to the variation in yield, meaning that variation in output from frontier is attributed to technical inefficiency and only 6 per cent of it was due to random factors outside their control.

The value of λ was more than one, implying the dominance of one sided component U_i in E_i and thus indicated high degree of technical inefficiency. In other words the inefficiency component was not dominated by the random factors only. It was found that, farmers in the study area had scope to increase maize productivity by attaining full efficiency through reallocating their resources. The return to scale (RTS) for the production function revealed that the farmers operated in the irrational zone (stage I) of the production surface having RTS of 1.092, which is greater than one indicating increasing returns to scale. This indicates that farmers in general, allocated their resources in the irrational zone of production (stage I) where the increasing returns to scale prevails, indicating if all the inputs specified in the function were increased by one per cent, output would have increased by 1.092 per cent by efficient use of resources allocated and used on their farms.

5.2.3 Efficiency Level of Sample Farms

The estimation of technical efficiency of the farmers revealed that the minimum technical efficiency of farmers was 8 per cent. The highest percentage of farmers came under the efficiency level of below 40 per cent and mean technical efficiency was 52 per cent, which revealed that on an average the sample farms operated 48 per cent below the frontier output levels. Hence, the maize crop output can be increased to 48 per cent by adopting proper technology.

5.2.4 Estimated Potential Yield

A significant variation was observed in potential yield and actual yield in different efficiency levels in sample areas. From this study, the most efficient farms (above 90 per cent of efficiency level) yield 14.88 qtls/ha and their potential yield was 25.64 qtls/ha for maize crop and the least efficient farmers having efficiency level of less than 40 per cent, produces 7.53 qtls/ha of maize with potential yield up to 13.98 qtls/ha. However, the most efficient and least efficient maize producers coexist in the same environment in a given region. The difference in technical efficiency was due to existing intra-farm efficiency within a region, which is due to difference in their level of input use.

5.2.5 Input Use and Technical Efficiency

One of the marked differences in determining inter-farm difference in technical efficiency is input use. From this study, the most efficient farms yielded up to 14.88 qtls/ha maize. In order to achieve this yield the inputs mainly the fertilizer dose of 46.44 N Kg/ha, 11.56 P Kg/ha and 12.20 K Kg/ha, 32.29Kg/ha of seed and 40.78 mandays/ha are to be used. The least efficient yield averages were 7.53 qtls/ha, which is obtained by applying the fertilizer up to 23.08 N Kg/ha, 7.01 P Kg/ha and 4.79 K Kg/ha, 20.58 Kg/ha of seed and 50.12 mandays/ha. Use of more labour by the least efficient producers is an evidence of substitution of labour as a factor of production. Labour is substituted for fertilizer because it is readily available. These differences can be removed by providing different inputs through government agencies on the basis of subsidies to the farmers in study area.

5.2.6 Factors Affecting Technical Efficiency

The efficiency determinants of maize crop likely to be affected by the same demographic and socioeconomic variables. The results of the multiple regression analysis indicate strong evidence that age of the farmer affects the technical efficiency. The younger farmers are more efficient than the older ones. These results showed that the older farmers were not adopting the new technology for maize crop production and not only have less contact with extension agents but also are less willing to adopt new practices and modern inputs for maize production in the study area. Based on the findings of this study, education had positive impact on the technical efficiency in maize production and technically educated farmers were more efficient. However, since the benefits of education are not instantaneous, the government should consider focusing on educating current farmers in best production practices. Non-formal agricultural education, often provided by both public and private extension services, is needed to provide the technical know-how to the farmers, farm families and workers of rural organizations and groups, which ultimately lead to increase the maize crop production. The coefficients for the proportion of female workers under study indicated that females were carrying out most crop management operations on the farm and thereby are contributing significantly to raise maize production levels. Besides, positive and significant values of proportion of children in a household indicated increasing level of technical inefficiency, because large family size put additional pressure on farm income for clothing, education, food and health care. In this study the farm size also showed the significant and positive relationship with regard to technical efficiency and it was concluded that farmer with fewer number of scattered holdings is more efficient than a farmer have large number of scattered fields. These findings imply relevant policy implications to raise productivity in study area. Loosening the various constraints that farm household face could possibly enable them to achieve a higher level of technical efficiency.

5.3 Economics (Costs and Returns) on Maize Production

It was concluded that the total overall average cost of cultivation of maize crop production in Doda and Udhampur district was ₹12882.83/ha and ₹10430.84/ha, respectively. It showed the decreasing trend in various costs with increase in the farm size, which was due to the higher expenditure on different variable inputs used by the marginal farmers as compared to large farmers, because of intensive farming system. Among the different inputs used in operational cost, overall average cost of human labour was the highest cost and contributed 32.54 per cent for Doda district and 36.61 per cent for Udhampur district to the total cost and it was observed that higher family labour utilization was done by marginal size farms than other category of farmers. Among the fixed cost overall average rental value of owned land was highest for both the districts and accounted for 35.56 per cent and 29.06 per cent to the total cost of cultivation.

It was also observed that the gross income as well as yield was more for small farmers as compared to large farmers for maize production in Doda and Udhampur district and overall average gross income was ₹17500.27/ha and ₹14276.85/ha, respectively. The reason for this was better management of small farms as compared to the large farms in study area. The all costs *i.e.*, A_1 and $A_{2,}$ $B_{1,}$ $B_{2,}$ $C_{1,}$ C_2 and cost C_3 was highest for marginal farmers and lowest for large farmers in both the districts but the benefit cost ratio was highest for small farmers and lowest for medium farmers in Doda district and in case of Udhampur district benefit cost ratio was highest for marginal farmers and lowest for large farmers.

5.4 Conclusion

It was concluded that growth rates in terms of area, production and yield was positive and significant during overall period at national and state level and in Jammu region the growth trend for yield was negative, where as in sample districts yield and production showed negative and significant growth trends. It was also concluded that only 11 per cent of the farmers came under the category of high level of technology adoption, maximum number of farmers used their farm produced seed and allocated maximum land under the traditional varieties as compared to hybrid varieties.

Among the different variables under study human labour, capital and fertilizers (N and K) were positively significant and could increase the yield by employing more units of these inputs in producing the maize crop in selected districts. Overall, mean technical efficiency of 52 per cent was achieved by maize farms in the study area. This shows there is a scope for increasing maize production by 48 per cent with the present technology itself and value of γ was 0.94, indicated that about 94 per cent of the differences between the observed and the maximum production frontier outputs were due to the factors, which were under farmers' control rather than random factors. This study provides the evidence that age of the farmer, education, female workers and the size of holding were the significant variables for improving technical efficiency among the sample farmers, where as the male workers and children in the family showed the negative relationship with technical efficiency under the study area.

5.5 Policy Implications

The maize scenario in future has to be seen in the light of competitiveness from other crops and measures have to be taken up to keep it profitable. The district having low yield have to be focused strongly to improve the yield. The technological adoption index is very poor and 49 per cent of the farmers in study area were low technology adopters. High costs involved inputs mainly, quality maize seeds, fertilizers, pesticides and micronutrients needs attention. Hence, financial assistance has to be given to farmers to purchase inputs. Furthermore, a government agency may be set up to procure the crop produce or any other institutional arrangement like contract farming may be initiated to assure the farmers about the marketability of their produce. The technical efficiency indicated that there is about 48 per cent potential for increasing the gross income of the farmers with existing levels of farmers' resources and technology, if the gap between the actual and potential yield is narrowed, the production frontier may be pushed further outward. This will help to increase not only the national pool of maize production but also increase the farm income of the maize growers considerably.

Bibliography

Abdulai, A. and Eberlin, R. 2001. Technical efficiency during economic reform in Nicaragua: Evidence from farm household survey data. *Economic System*, **25**(2): 113-125.

Adeleke, O.A. Fabiyi, Y.L. and Matanmi, H.M. 2008. Application of Stochastic Frontier in the estimation of technical efficiency of Cassava farmers in Oluyole and Akinyele local government areas of Oyo State. *Research Journal of Agronomy*, **2** (3): 71-77.

Ahmad, M., Chaudhry, G.M. and Iqbal, M., 2002. *Wheat productivity, efficiency, and sustainability: A stochastic production frontier analysis. The Pakistan Development Review*, **41** (4): 643-663.

Aigner, D.J., C.A.K. Lowell and P. Schmidt, 1977. Formulation and estimation of stochastic frontier production function models. *J. Economics*, **6**(1): 21-37.

Ajao, A.O., Ajetomobi, J.O. and Olarinde, L.O. 2005. Comparative efficiency of mechanized and non-mechanized farms in Oyo State of Nigeria: A stochastic frontier approach. *Journal of Hum. Ecol.*, **18** (1): 27-30.

Alam, Q. M. 1989. Resource use efficiency of irrigated and non-irrigated farms in a selected area of Bangladesh. *Economic affairs*, **34** (4): 271-277.

Ali, H.E., Balbase, K., Grabowski, R. and Kraft, S. 1987. The technical efficiency of Illinois grain farms: An application of a ray-homothetic production function. *Southern Journal of Agricultural Economics,* **19**: 69-78.

Ali, M. and Choudhry, M.A. 1990. Inter-regional farm efficiency in Pakistan's Punjab: A frontier production function study. *Journal of Agricultural Economics,* **41** (1): 62-74.

Ali, M. and Flinn, J.C. 1989. Profit efficiency among basmati rice producers in Pakistan's Punjab. *American Journal of Agricultural Economics,* **17**: 303-310.

Amaza, P. S. and Olayemi, J. K. 2000. Technical efficiency in food crop production in Gambe State, Nigeria. *Niger Agricultural Journal,* **32**:140-151. Anonymous 2008. Digest of Statistics. Directorate of Economics and Statistics, Planning and Development Department, Govt. of Jammu and Kashmir.

Anonymous, 2008. Government of India, Economic Survey, Economic Division, Ministry of Finance, New Delhi.

Anonymous 2008. Digest of Statistics. Directorate of Economics and Statistics, Planning and Development Department, Govt. of Jammu and Kashmir.

Anupama, J. Singh, R.P. and Kumar, R. 2005. Technical efficiency in maize production in M.P. *Agricultural Economics Research Review,* **18:** 305-315.

Arya, S.L. and Rawat, B.S. 1990. Agricultural growth in Haryana. A district-wise analysis. *Agricultural Situation in India,* **45** (2): 33.

Atibudhi, H.N. 2003. Maize production in Orissa: District level analysis. In : Maize Production in India- Golden Grain in Transition (Eds. Kumar, R. and Singh, N.P.).Technical bulletin TB- ICN: 4/2003, Division of Agricultural Economics, Indian Agricultural Research Institute, New Delhi, pp. 77-87.

Audibert, M., 1997. Technical efficiency effects among paddy farmers in the villages of the Office du Niger Mali, West Africa. *J. Productivity Anal.,* **8**(4): 379-394.

Awasthi, P.K. 2003. Maize production in Madhya Pradesh: A district level analysis. In : Maize Production in India- Golden Grain in Transition (Eds. Kumar, R. and Singh, N.P.).Technical bulletin TB- ICN: 4/2003, Division of Agricultural Economics, Indian Agricultural Research Institute, New Delhi, pp. 60-76.

Awudu, Abdulai and Wallace Huffman. 2000. Structure adjustment and economic efficiency of rice farmers in Northern Ghana. *Economic Development and Cultural Change*, 503-519.

Awudu, Abdulai and Richard Eberlin. 2001. Technical efficiency during economic reform in Nicaragua: Evidence from farm household survey data. *Economic Systems*, **25**: 113-125.

Badal, P. S. and Singh, R. P. 2001. Technological change in maize production: A case study of Bihar. *Indian Journal of Agricultural Economics*, **56:** 211-19.

Battese, G.E. and Coelli, T.J. 1995. A model for technical inefficiency effects in a stochastic frontier production function for panel data. *Empirical Economics*, **20:** 325-332.

Bedassa, Tadesse and Krishnamoorthy, S. 1997. Technical efficiency in paddy farms of Tamil Nadu: An analysis based on farm size and ecological zone. *Agricultural Economics*, **16**:185-192.

Bravo-Ureta, B. and Pinheiro, A. 1993. Efficiency analysis of developing countries agriculture: A review of the frontier function literature. *Agriculture and Resource Economics Review*, **22** (1): 88-101.

Bravo-Ureta, B.E. and Evenson, R.E. 1994. Efficiency in agricultural production: the case of peasant farmers in eastern Paraguay. *Agricultural Economics*, **10**(1): 27-37.

Chahal, S.S. and Kataria, P. 2003. Maize production in Rajasthan: A district level analysis. In : Maize Production in India- Golden Grain in Transition (Eds. Kumar, R. and Singh, N.P.).Technical bulletin TB-ICN: 4/2003, Division of Agricultural Economics, Indian Agricultural Research Institute, New Delhi, pp 88-105.

Chand,Ramesh and Raju, S.S. 2008. Instability in Andhra Pradesh agriculture- A disaggregate analysis. *Agricultural Economics Research Review*, **21**: 283-288.

Chandra, N. and Singh, R.P. 1992. Determinant and impact of new technology adoption on tribal agriculture in Bihar. *Indian Journal of Agricultural Economics*, **47** (3): 397-403.

Chowdhury, J. M. 1992. Study on production potential and economic evaluation of maize based cropping systems under mid-hill conditions of Himachal Pradesh. *Farming Systems*, **8**(3/4): pp 70-78.

Coelli, T. and Battese, G. 1996. Identification of factors which influence the technical inefficiency of Indian farmers. *Australian Journal of Agricultural Economics*, **40**(2): 103-128.

Coelli, T.J., Prasada, D.S. and Battese, G.E. 1998. An introduction to efficiency and productivity analysis.Kulwer Academic Publishers Boston Dordrecht, pp 108-113.

Dawson, P.J. and Lingard, J. 1989. Measuring farm efficiency over time Philippine rice farm. *Journal of Agricultural Economics*, **40** (2): 168-176.

Dey, M.M., Paraguas, F.J., Bimbao, G.B. and Regaspi, P.B. 2000. Technical efficiency of tilapia grows out pond operations in the Philippines. *Aquacultural Economic Management*, **4** (1-2): 33-47.

Dolisca, F. and Jolly C. M. 2008. Technical efficiency of traditional and non-traditional crop production: A Case Study from Haiti. *World Journal of Agricultural Sciences*, **4** (4): 416-426.

Duloy, J. H. 1959. Resource allocation and fitted production function. *Australian Journal of Agricultural Economics*, **3**(2): 33-39.

Elsamma, J. and George, M.V. 2002. Technical efficiency in rice production- A frontier production approach. *Agricultural Economic Research Review*, **15** (1): 50 - 55.

Farrel, M.J. 1957. The measurement of productive efficiency. *Journal of Royal Statistics Society*, **120** (4): 253-290.

Forsund, F.R., Lovell, C.A.K. and Schmidt, P. 1980. A survey of frontier production function and of their relationship to efficiency measurement. *Journal of Econometrics*, **13**: 5-25.

Ghaderzadeh, H. and Rahimi, M.H. 2008. Estimation of technical efficiency of wheat farms. A case study in Kurdistan province, Iran. *American Eurasian Journal of Agriculture and Environmental Sciences*, **4**(1): 104-109.

Goni, M., Mohammed, S. and Baba, B. A. 2007. Analysis of resource-use efficiency in rice production in the lake Chad area of Borno State, Nigeria. *Journal of Sustainable Development in Agriculture and Environment,* **3**:31-37.

Goyal, S.K. and Suhag K.S. 2003. Estimation of technical efficiency on wheat farms in Northern India-A Panel Data Analysis. *International Farm Management Congress*, pp-1-14.

Greene, W.H. 1980. Maximum likelihood estimation of econometric frontier functions. *Journal of Econometrics*, **13**: 27-56.

Hasan, F.M. 2008. Economic efficiency and constraints of maize production in the northern region of Bangladesh. *J. Innov.Dev.Strategy*, 2(1): 18-32.

Hazarika, C. and Subramanian, S.R. 1999. Estimation of technical efficiency in the stochastic frontier production function model –An application to the tea industry in Assam. *Indian Journal of Agricultural Economics*, **54** (2): 201–211.

Hussain, S.S., 1989. Analysis of economic efficiency in Northern Pakistan: Estimation, causes and policy implications. PhD. Dissertation, University of Illinois.

Idiong, I.C. 2005. Estimation of farm level technical efficiency in small scale swamp rice production in cross river state of Nigeria: A stochastic frontier approach. *World Journal of Agricultural Sciences*, **3** (5): 653-658.

Idiong, I.C., Onyenweaku, E., Susan Ohen, B. and Damian Agom, I. 2007.A stochastic frontier analysis of technical efficiency in swamp and upland rice production systems in cross river state, Nigeria. *Agricultural Journal*, **2** (2): 299-305.

Ingosi, A. 2005. Economic evaluation of factor influencing Maize yield in the North Rift region of Kenya. Masters of Science Thesis, Colorado State University

Javed, M.I., Adil, S.A., Javed, M.S. and Hassan, S. 2008. Efficiency analysis of rice-wheat system in Punjab, Pakistan. *Pakistan Journal of Agricultural Sciences*, **45** (3): 95-100.

Jayadevan, C.M. 1991. Instability in wheat production in Madhya Pradesh. *Agricultural Situation in India*. **XLVI** (4): 219-223.

Jha, B. K. 1996. Resource use efficiency on mixed farms in the green belt of India. *Agricultural Situation in India,* **51** (8): 559-564.

Jyothirmai, U. L., Shareef, S. M. and Panduranga Rao, A. 2003. Resource use efficiency in three regions of Nagarjuna Sagar command area in Andhra Pradesh. *Economic affairs,* **48** (1): 59-65.

Kachroo, Jyoti and Kachroo, Dileep. 2006. Growth, instability and projection of silk production in Jammu and Kashmir State. *Journal of Research,* SKUAST-Jammu, **5**(1): 20-34.

Kachroo, J. and Kachroo, D. 2007. Economic analysis of fine rice production under subtropical agro climatic zone of Jammu region of Jammu and Kashmir State. *Agricultural Situation in India,* **LXIV** (9): 413-17.

Kachroo, J., Sharma, A. and Singh, T. 2008. Technical efficiency of paddy crop in Jammu district of J&K State. *Agricultural Situation in India,* **LXVI** (3):151-155.

Kachroo, J. and Sharma, M. 2008. Sunflower scenario in India. Journal of Research, SKUAST-Jammu, **7**(1): 122-131.

Kalirajan, K.P. 1981. An econometric analysis of yield variability in paddy production. *Canadian Journal of Agricultural Economics,* **29**: 283-294.

Kalirajan, K.P. and Flinn, J.C. 1983. The measurement of farm-specific technical efficiency. *Pakistan Journal of Applied Economics,* **2**:167-180.

Kalirajan, K.P. and Shand, R.T. 1986. Estimating location- specific and firm –specific technical efficiency: An analysis of Malaysian agriculture. *Journal of Economic Development,* **11**: 147-160.

Kalirajan, K.P. and Shand, R.T. 1994. Modelling and measuring technical efficiency. An alternative approach. In issues in agricultural competitiveness: Markets and policies, I.A.E.E, *Occasional Research Paper No.7,* pp-232-239.

Kamruzzaman, M. and Hedayetul, M.I. 2008. Technical efficiency of wheat growers in some selected sites of Dinajpur district of Bangladesh. *Bangladesh Journal of Agricultural Research,* **33** (3): 363-373.

Khan, D., Bashir, M. and Zulafkar, M. 2005. Estimation of net returns from main crops in district Malakand. *Journal of Applied Sciences*, **5** (9): 1564-68.

Khosta, A.K. and Chandrakar, M.R. 2005. A comparative study of economic efficiency in production of irrigated and rainfed rice in Chattisgarh. *Indian Journal of Agricultural Economics*, **60** (3): 524.

Kibaara, B.W. 2005. Technical efficiency in Kenyan's maize production: An application of the Stochastic Frontier Approach, Department of Agricultural and Resource Economics, Colorado State University Fort Collins, Colorado.

Kopp, R.J and Smith, V.K. 1980. Frontier production function estimations for steam electric generation: A comparative Analysis. *Southern Econometric Journal*, **47**: 1049-59.

Kudi, T.M. and Abdulsalam, Z. 2008. Costs and return analysis of Striga tolerant maize variety in Southern Guinea Savana of Nigeria. *Journal of Applied Sciences Research*, **4** (6): 649-651.

Kumar, N. 1995. Economics of different rice based/competing crop sequences of Punjab, M.Sc. thesis, PAU, Ludhiana.

Kumar, R. Singh, N.P. and Vasisht, A.K. 2004. Adoption pattern of improved maize technology in Northern India: Impact on farm earning and trade. *Agricultural Economics Research Review*, **17**: 29-42.

Kumar Sikander and Kumar Sandeep. 2004. Resource use efficiency and returns from selected foodgrain crops of Himachal Pradesh: A study of low hill zone. *Agricultural Situation in India*, **59** (7): 475-485.

Kumar, L.R., Srinivas, K. and Singh, S.R.K. 2005. Technical efficiency of rice farms under irrigated conditions of North West Himalayan region-a non-parametric approach. *Indian Journal of Agricultural Economics*, **60**(3):483-492.

Kumbhakar, C. Subbal. 1994. Efficiency estimation in a profit maximizing model using flexible production function. *Agricultural Economics*, **10**: 143-152.

Lingard, J., Castillo, L. and Jayasuriya, S. (1983). Comparative efficiency of rice farms in central Luzon. *The Philippines Journal of Agricultural Economics*, **34:** 37-76.

Liu Zinan and Zhuang Juzhong, 2000. Determinants of technical efficiency in post-collective Chinese agriculture: Evidence from farm-level data. *Journal of Comparative Economics*, **28**: 545-564.

Meeusen, W. and Van Den Broeck, J. 1977. Efficiency estimation from Cobb-Douglas production functions with composed errors. *International Economic Review*, **18**(3): 163-173.

Mohiuddin, M., Karim, M.R., Rashid, M.H. and Hudda, M.S. 2007. Efficiency and sustainability of maize cultivation in an area of Bangladesh. *Int. J. Sustain. Crop Production.* **2** (3):44-52.

Mrigendra, S. and Neelkanth, P. 1996. Economic analysis of crop production in M.P. during 1969-70 to 1991-92. *Agricultural Situation in India,* **53** (2): 89-92.

Neeraj, R., Chanda, S., Srivastava, S.C. and Singh, P.K. 2003. An economic analysis of wheat production on different sizes of farm in Kanpur (Dehat) district of Uttar Pradesh. *The Bihar Journal of Agricultural Marketing,* **11** (384): 159-166.

Ogundari, K. 2006. Resource-productivity, allocative efficiency and determinants of technical efficiency of rainfed rice farmers: A guide for food security policy in Nigeria. *Journal of Sustainable Development in Agriculture and Environment,* **3** (2):20-33.

Ogundari, K. and Ojo, S. O. 2007. Economic efficiency of small scale food crop production in Nigeria: A Stochastic Frontier Approach. *Journal of Social Sciences,* **14**(2): 123-130.

Oluwatayo, I.B. Sekumade, A.B. and Adesoji, S.A. 2008. Resource use efficiency of maize farmers in rural Nigeria: Evidences from Ekiti State. *World Journal of Agricultural Sciences,* **4** (1): 91-99.

Onyenweaku, C.E. and E.O. Effiong, 2005. Technical efficiency in pig production in Akwalbom State, Nigeria. *International Journal of Agricultural Rural Development,* **6**: 51-57.

Oyewo, I.O., Rauf, M.O., Ogunwole, F. and Balogun, S.O. 2009. Determinants of maize production among maize farmers in Ogbomoso south local government in Oyo State. *Agricultural Journal,* **4** (3): 144-149.

Radha, Y. and Prasad, Y. E. 1999. Variability and instability of area, production and productivity of rice and maize in Northern Telangana zone of Andhra Pradesh. *Agricultural Situation in India,* **53** (10): 623-626.

Raghuwanshi, R.S., Awasthi, P.K. and Sharma, P. 1999. Resource use efficiency in wheat cultivation. *Indian Journal of Agricultural Research,* **33** (1): 67-71.

Rajshekharan, P. and Krishnamoorthy, S. 1998. Technical efficiency and pesticide use in rice production. *Agricultural Economics Research Review,* **2** (1): 1-9.

Sabur, S. A. and Haque, M. Z. 1992. Resources use efficiency and returns from some selected winter crops in Bangladesh. *Economic Affairs,* **37** (3): 158-68.

Sahota, G. S. 1968. Efficiency of resource allocation in Indian agriculture. *American Journal of Agricultural Sciences, **50*** (3): 589-605.

Salik, Ram and Gupta, S. B. Lal 1978. Resource productivity on paddy farms in Chandauli block of Varanasi district. *Agricultural Situation in India,* **33** (6): 373-374.

Seyoum, E.T., Battese, G.E. and Fleming, E.M. 2000. Technical efficiency and productivity of maize producers in Eastern Ethiopia: A study of farmers within and outside the Sawakawa–Global 2000 Project. *Agricultural Economics,* **19**: 341-348.

Shanmugam, K.R. and Palanisami, K. 1993. Measurement of economic efficiency-Frontier function approach. *Journal of Indian Society of Agricultural Statistics,* **45** (2): 235-242.

Shanmugam, K.R. 2001. A comparative study of technical efficiency of farms in Bihar, Karnataka, Punjab and Tamil Naidu: A random coefficient approach. In accelerating growth through globalization of Indian agriculture (Eds. K.P. Kalirajan, G. Mythilli and U. Shanker). Machmillan India Limited, New Delhi, pp. 288-323.

Shanmugan, K.R. 2003. Technical efficiency of rice, groundnut and cotton farms in Tamil Naidu. *Indian Journal of Agricultural Economics,* **58** (1): 101-104.

Shanmugm, K.R., and Venkataramani, A. 2006. Technical efficiency in agricultural production and its determinants. An exploratory study at district level. *Indian journal of agricultural Economics*, **16**(2): pp-169-184.

Sharma Ravinder and Moorti, T. V. 1989. Economics of Resource use: A study of Himachal tea farms. *Agricultural Situation in India,* **44**(2): 117-120.

Sharma, V. K., Kingra, H. S. and Singh, J. 2003. An economic analysis of Basmati vis-à-vis non-Basmati rice in Jammu and Kashmir. *Indian Economic Panorama,* **13**: 46-48.

Shehu, J.F. and Mshelia, S.I. 2007. Productivity and technical efficiency of small scale rice farmers in Adamawa state, Nigeria. *Journal of Agriculture and Social Sciences,* **3** (4): 117-120.

Singh, J. P. 2005. An economic analysis and resource use efficiency of canal irrigated paddy vis-à-vis tubewell irrigated paddy in Faizabad district of eastern U.P. *Indian Journal of Agricultural Economics,* **60** (3): 537.

Singh, K. 1971. Resource adjustment possibility and resource use efficiency in Punjab agriculture. Ph.D. dissertation Punjab Agricultural University, Ludhiana, India.

Singh, Karam, Rangi, P.S. and Kalara Sajala. 2004. Wheat production and sustainability in Punjab: growth and varietal diversity. *Indian Journal of Agricultural Economics,* **59** (4): 745-771.

Singh, N., Sandhu, H. S. and Singla, S. K. 1992. Resource use efficiency of irrigated and unirrigated wheat in kandi region of Punjab. *Indian Journal of Agricultural Economics,* **47:** 555-56.

Singh, R. P. 2001. Study of resource use efficiency in maize and ragi production in plateau region, Bihar. *Economic Affairs,* **46** (4): 222-226.

Singh, R.P., Kumar, R.and N.P. Singh, 2003. Transformation of the Indian maize Economy- Different Perspectives. In : Maize Production in India- Golden Grain in Transition (Eds. Kumar, R. and Singh, N.P.).Technical bulletin TB- ICN: 4/2003, Division of Agricultural Economics, Indian Agricultural Research Institute, New Delhi, pp. 1-28.

Singh, S. 1999. A study on technical efficiency of wheat cultivation in Haryana. *Agricultural Economics Research Review,* **20**: 127-136.

Singh, S.P., Gangawar, B. and Garg, B.M. 2007. Input use efficiency- A comparative study of rice –wheat and sugarcane ratoon cropping system in mid Western plain of U.P. *Agricultural Situation in India,* **LXVIII** (3): 377-383.

Suresh, A. and Reddy, T. R. Keshav 2006. Resource use efficiency of paddy cultivation in Peechi command area of Thrissur district of Kerela: An economic analysis. *Agricultural Economics Research Review,* **19**: 159-171.

Verghese, K.A. and Rathore, R.S. 2003. Maize production in Rajasthan: A district level analysis. In : Maize Production in India- Golden Grain in Transition (Eds. Kumar, R. and Singh, N.P.).Technical bulletin TB-ICN: 4/2003, Division of Agricultural Economics, Indian Agricultural Research Institute, New Delhi, pp. 106-120.

Wadud, A. and White, B. 2000. Farm household efficiency in Bangladesh: A comparison of stochastic frontier and DEA methods. *Journal of Applied Economics,* **32**(13): 1665-1673.

Yotopoulos, P. A. 1967. Allocative Efficiency in Economic Development, Research Monograph Series, No. **18** (191-192), Constantinidis and C. Mihalas, Athens.

Yotopoulos, P. A. and Nugent, G.B. 1976. Economics of development, empirical investigation. Harper and Row Publishers, 10 East 53rd Street New York, N.K. 10082, pp. 76.